T0260983

Cattle Country

AT TABLE

CATTLE COUNTRY

LIVESTOCK *in the* CULTURAL IMAGINATION

Kathryn Cornell Dolan

University of Nebraska Press | Lincoln

Portions of chapter 3 first appeared as "Eating Moose:
Thoreau, Regional Cuisine, and National Identity"
in *Rediscovering the Maine Woods: Thoreau's Legacy
in an Unsettled Land*, ed. John J. Kucich (Amherst:
University of Massachusetts Press, 2019), 123–38.

Chapter 4 first appeared as "Cattle and Sovereignty in
the Work of Sarah Winnemucca Hopkins," *American
Indian Quarterly* 44, no. 1 (Winter 2020): 86–114.

Publication of this volume was assisted by the Virginia
Faulkner Fund, established in memory of Virginia Faulkner,
editor in chief of the University of Nebraska Press.

Library of Congress Cataloging-in-Publication Data
Names: Dolan, Kathryn Cornell, author.
Title: Cattle country: livestock in the cultural
imagination / Kathryn Cornell Dolan.
Description: Lincoln: University of Nebraska Press, [2021] |
Series: At table | Includes bibliographical references and index.
Identifiers: LCCN 2020038663
ISBN 9781496218643 (hardback)
ISBN 9781496226990 (epub)
ISBN 9781496227003 (mobi)
ISBN 9781496227010 (pdf)
Subjects: LCSH: Cattle in literature—United States. | Cattle—
United States—History. | Livestock—United States—History.
Classification: LCC SF196.U5 D65 2021 | DDC 636.2—dc23
LC record available at https://lccn.loc.gov/2020038663

Set in Scala OT by Laura Buis.
Designed by L. Auten.

To the Dolans, with laughs and lots of love

CONTENTS

ILLUSTRATIONS

ACKNOWLEDGMENTS

First, I'd like to thank all my friends and colleagues at Missouri University of Science & Technology. For their teamwork and friendship, I'd like to thank the Department of English and Technical Communication, especially Kristine Swenson and Linda Sands. Thanks also to the College of Arts, Sciences, and Business dean's office and provost's office—Stephen Roberts, Kate Drowne, Shannon Fogg, Melanie Mormile, Yinfa Ma, Kristen Gallagher, and Linda Bramel—for everything you've done to help me professionally and personally during my time at Missouri S&T. This book would not be what it is without the opportunities you've provided. My Saturday work group—Carleigh Davis, Sarah Hercula, and Rachel Schneider—you all are the best—and we really need a nickname, don't we? My walking buddy, Anne Cotterill: I'm healthier and happier because of you. My traveling companion colleagues, Kathryn Northcut, Audra Merfeld-Langston, and Susan Murray: you survived the greatest challenge, namely, me. John Morgan, who read early drafts of several chapters and gave invaluable feedback: thanks for your American literature experience—my chapters are the better for your eye. My book club and happy hour crew—Mike and Jeanine Bruening, Andrew and Lucy Behrendt, Justin Pope and Alanna Krolikowski, Agnes Vojta, Pat Huber, Dan and Beth Reardon, John Gallagher, Matt Thimgan, Katie Shannon, Jill Ho, Cate

Johnson, Jessica Cundiff, Kate Sheppard, and Amber Henslee—here's to many more gatherings at our houses and The Public House. (Special mention goes to honorary happy hour participant Simon Bronner.) My book club friends—Margie Edberg, Leslie Bearden, Suzanne Femmer, Phyllis Meagher, Patti Fleck, Cynthia Riley, Lina Madison, Dianna Morrison, Gayla Olbricht, Havva Malone, Jane Schonberg, Janet Mckean, Le Wang, Marg Olson, Casey Wilson, Pat Oster, Catherine Tipton, and Sheila Packard—thanks for our monthly gatherings that encourage me to read something outside of the nineteenth century!

In addition, I am grateful for the financial support I received from Missouri s&t and the um system, which gave me much-needed teaching releases at critical points in this book's development. A 2016 University of Missouri Research Board grant allowed me time to develop my second monograph during its early stages, and a 2019 Missouri s&t Sabbatical and 2020 University of Missouri Research Board grant allowed me to complete the manuscript. Teaching, of course, is the fun part, as well. I value all of my students at Missouri s&t, especially those in my early U.S. literature surveys and seminars, as well as the members of Sigma Tau Delta. I consistently am amazed by how much I learn from you, and you all make me a better teacher and researcher.

Outside of the Missouri s&t community, I have received valuable support on my work from reading groups, invited talks, institutes, and conferences. I'm delighted to be a new member of the Midwest Nineteenth-Century Americanists writing group, where I was able to workshop my introduction in 2019. Thanks especially to Melissa Homestead, Laura Mielke, and Juliana Chow for your timely comments. Jamille Palacios Rivera provided me with a productive opportunity to present work from this project at the Jill Findeis Seminar Series at the University of Missouri's College of Agriculture, Food, and Natural Resources in 2017. My thanks also to the audience there, who asked thoughtful questions and made this book all the stronger for their help. I was part of a true community of scholars at the 2016 NEH Summer Institute, "Extending the Land Ethic," where I got to work on a draft of the chapter on Charles Chesnutt and agriculture. I'm particularly grateful to Brittany Davis, Samantha Noll, Laura Hartman, Sebastian Purcell, Alexandria Poole, Duncan Purves, Katey Castellano, Todd Richardson, and the rest of the crew. Thanks to Dan Shilling and Joan McGregor, as well, for organizing a wonderful summer camp for scholars. The opportunities to present drafts of this work at a variety of conferences has been invaluable. Thanks to John Kucich at the American Literature Association conference (2015) for his invaluable feedback on what became the chapter on Thoreau, the Maine Woods, and Cape Cod. Thanks also go to Shanna Salinas, Carina Evans, and Sharon Lynette Jones for participating on panels with me at Midwest Modern Language Association (MMLA) conferences (2016, 2017, 2018), helping me revise material that would make up much of my work on Washington Irving in the Indian Territory, Sarah Winnemucca Hopkins's experiences with cattle ranches, and María Amparo Ruiz de Burton's descriptions of vaqueros. In addition, I'd like to thank my MMLA executive committee peers, Andrea Knutson, Erika Berisch Elce, Eloise Sureau, Michelle Medieros, Sheila Liming, Tisha Brooks, Justin Hastings, Emily Isaacson, Matthew Barbee, and especially Christopher Kendrick and Linda Winnard. I truly appreciated our work together, and I continue to learn valuable service lessons from you all. Thanks as well to Lauren LaFauci at the 2019 Association for the Study of Literature and Environment symposium, where I presented a portion of my chapter Washington Irving and the framing of the West in the early nineteenth century. Tom Hertweck, Lisa Swanstrom, and Laura Wright also provided good advice over food and drink. Food studies is definitely a field with benefits!

In addition, I'd like to thank my editor at University of Nebraska Press, Emily Wendell, as well as my reviewers and the advisory board at the press. I'm delighted to work with this press a second time. In addition, I received much appreciated support from two library archives in particular. Thanks to Jacquelyn K. Sundstrand at the Mathewson–IGT (International Game Technology) Knowledge Center, University of Nevada, Reno, for her assistance in accessing Sarah Winnemucca Hopkins's manuscripts and transcripts. Also, thanks to Erika Dowell and her associates at the Lilly Library, University of Indiana, Bloomington, for their help with the Upton Sinclair archive.

I'd be lost without my California homes away from home, where I've been able to make my biannual visits for the last decade. Wendy Norberg, Jennifer and Kuin Berkman and family, Sarah Hirsch, Rachel Mann, and Raidis Maypa: it's always a pleasure to spend time with you, and I've been beyond lucky to be able to visit you as often as I have. Finally, I want to again thank my crazy and loving family: Doug, Sandy, Kevin, John, Kristi, and Olivia. And last but not least, I want to extend another fond thank-you to my cats, Aussie and Missy, those beloved clichés of the academic writer. Keep trying your best to distract me as I write . . . it helps my focus!

Cattle Country

Fig. 1. This iconic image of nineteenth-century U.S. expansion, John Gast's *American Progress* (1872), also referred to as *Manifest Destiny*, depicts U.S. agriculture plowing away the people and animals on the painting. Courtesy of the Autry Museum of the American West.

Introduction

Cattle and Progress

On September 20, 1806, as they neared St. Charles, Missouri, at the end of their journey, William Clark, of the Lewis and Clark Expedition, noted in his journal, "We saw some cows on the bank which was a joyfull Sight to the party and caused a Shout to be raised for joy."[1] For Clark, the return to cattle was the return to civilization, so much so that the men of the party actually exclaimed in pleasure. Clark writes "joy" repeatedly in this sentence. This intensity of emotion regarding cattle is telling at the beginning of the nineteenth century. In fact, cattle make a useful metaphor for the civilizing idea of the Euro-American settler throughout the nineteenth century. Clark begins the century discussing expansion connected with agriculture, what I call agri-expansion, into the West, and this theme continues throughout the long nineteenth century with other writers and artists. In a famous painting of 1872, *American Progress*, John Gast presents a golden-haired female figure symbolizing "Progress" moving from the east to west across the canvas (fig. 1). The painting is dense with symbols representing U.S. ideas of progress, such as telegraph wires, trains, and steamships. As one of his signs of progress, Gast prominently shows the spread of U.S. agriculture; the scene regarding farming makes up a significant portion of the painting. Oxen plow the field at the bottom right, seemingly closest to the viewer; meanwhile, toward the top left, bison are being ridden off the canvas.[2] In the narrative of this work, cattle are almost directly forcing the bison off of the prairies and out of sight, to make room for themselves and their human companions. Meanwhile, U.S. farmers—and to a lesser extent, hunters—are maneuvering the indigenous communities of the Great Plains off the canvas, as well, moving almost as steadily westward as lady "Progress" is. While this is merely one image in a painting imbued with elements of nineteenth-century Manifest Destiny iconography, this contrast between two forms of cattle, domestic cows and indigenous bison—the U.S. form of cattle production and other forms of engagement with the environment—is the focus of this book.

Cattle Country: Livestock in the Cultural Imagination developed directly from my first monograph, *Beyond the Fruited Plain: Food and Agriculture in U.S. Literature, 1850–1905* (2014). While that book analyzes national ideas of expansion and foodstuffs in nineteenth-century U.S. literature, it does not directly examine what I have come to realize is the representative U.S. livestock animal, the cow. Appreciating this, I more deeply consider the role of cattle in U.S. regional narratives during the nineteenth century in *Cattle Country*. The drive for beef, the nation's striking demand for cheap and plentiful meat, was directly connected to the desire for territorial expansion—for lands on which to raise cattle. This evolution of U.S. cattle production in the nineteenth century came at a high cost to humans, animals, and the land, and these costs were recognized at the time by the period's authors and other artists. For example, the stories being told during the nineteenth century, especially in its first half, were influential in driving agri-expansion. By the second half of the century, though, authors became more openly critical of westward movement related to livestock production. They proposed alternatives to or challenged the development of intensifying agriculture. Observing these themes in nineteenth-century narratives, I developed a sustained study of cattle as an image in the work of several nineteenth-century authors. In *Cattle Country*, therefore, I argue that my chosen authors—Washington Irving, James and Susan Fenimore Cooper, Henry David Thoreau, Sarah Winnemucca Hopkins, María Amparo Ruiz de Burton, Charles Chesnutt, Upton Sinclair, and Winnifred Eaton—use cattle in their writings to investigate (either colluding with or criticizing) developing U.S. agri-expansion that mirrored the nation's advancement over the North American continent.[3]

Certainly, animal husbandry is one of the key elements of agriculture. When the Europeans arrived in the Americas, they brought their foods—including domestic animals—with them. Cattle came to the region that would become the United States with European colonialism, along with other domestic animals like pigs, chickens, horses, and, to a lesser extent, sheep and goats. Virginia DeJohn Anderson historicizes livestock as a colonizing tool in North America, reasoning that the establishment of populations of cattle and other domestic animals was a key project for early British colonists to the continent and soon also with the indigenous nations and tribes as they sought to navigate their changing world.[4] Even as early as the seventeenth century, livestock were inexorably linked to Euro-American expansion, taking on a commodity status as part of a global economics. As New Englanders turned cheap grazing lands into profit through beef cattle, Caribbean colonizers were able to focus on increasing their profits in the production of sugar—perhaps the largest global industry of the time. This focus on livestock and trade, relying on ostensibly cheap North American grazing lands, continued into the nineteenth century. At that time, the nation sought to add land to its borders through virtually any means necessary: the Louisiana Purchase of 1803, the Indian Removal Act of 1830, the Mexican-American War ending with the Treaty of Guadalupe Hidalgo in 1848, the acquisition of Alaska in 1867, the Spanish-

American War and annexation of Hawaii in 1898, and the continuing Indian Wars throughout the century resulted in vastly larger physical borders for the United States by the end of the nineteenth century. The motivation for these imperialistic actions was the nation's desire for greater expanses of territory—often bloodily acquired—that could be turned into agricultural lands, largely connected to cattle ranches and farms growing cereal grains for feed. Cattle, non-indigenous animals, therefore, were able to become a dominant symbol over the course of early U.S. history, and they have remained so into the twenty-first century.[5]

In the early nineteenth century, though, as the United States was beginning its most physically expansionist period, cattle were practically nonexistent in the Great Plains region of the continent. In fact, Meriwether Lewis and William Clark make almost no note of cows in the journals they kept during their travels west of St. Louis on the Corps of Discovery Expedition, more familiarly known as the Lewis and Clark Expedition. Not only do they not witness these domestic animals to the west of the established lands of the United States; they also do not imagine domestic cattle in the place of wild bison in their journal entries in the way that Washington Irving and James Fenimore Cooper would do in their western writings. Instead of comparing bison to cattle, the members of the expedition compare bison to grizzly bears in size, to deer and elk in terms of usefulness as a foodstuff, and to beavers as harvestable commodities. As the party moves west of the Rocky Mountains, even bison fade from Lewis's list of wild animals. On February 15, 1806, Lewis writes: "The quadrupeds of this country from the Rocky Mountains to the pacific Ocean are 1st the *domestic animals*, consisting of the horse and dog only; 2edly the *native wild animals*," after which he provides a list including bear, deer, elk, wolf, beaver, otter, raccoon, and panther.[6] Neither form of cattle—bison nor domestic cow—was part of the animal husbandry in western North America in early

1806, the time of the expedition's return voyage. Rather, their journals provided U.S. expansionists with exploitative harvesting opportunities in the West, like beaver for their pelts. At this point, the western lands of the continent were being considered for their extractive economic potential, not as lands suitable for U.S. settlement.[7]

Clark does engage with agriculture, though, in his journal during a September 12, 1806, meeting with Arikara and Sioux interpreters returning from Washington DC. The Corps of Discovery Expedition had first encountered the Arikara in 1804. At that time, the expedition had encouraged a delegation to travel to Washington, where, in addition to the death of Chief Ankedoucharo (spelled Ar-ke-tar-na-Shar in his journal), another member, Jo Gravelin, was being sent back to "teach the Ricaras agriculture."[8]

We found Mr. Jo. Gravelin the Ricaras enterpreter whome we had Sent down with a Ricaras Chief in the Spring of 1805. and old M. Durion the Sieux enterpreter, we examined the instructions of those interpreters and found that Gravelin was ordered to the Ricaras with a Speech from the president of the U. States to that nation and some presents which had been given the Ricara Chief who had visited the U. States and unfortunately died at the City of Washington, he was instructed to teach the Ricaras agriculture & make ever enquirey after Capt. Lewis my self and the party.[9]

In this way, Clark implies that the general belief of U.S. policy makers was that the Arikara as a people lacked what was considered an appropriate, or Euro-American, knowledge of agriculture. As early as in Clark's 1806 journal entry, then, the work of compelling a particular system of agriculture on other nations on the continent was already apparent, and it would grow more prevalent and extreme throughout the nineteenth century. Since first contact, Native communities had been required to adapt to U.S. forms of agriculture—including animal husbandry—in order to be considered civilized and worthy of the land, as Reginald Horsman argues.[10] This pressure reached its apex during the nineteenth century, the era of westward expansion and Manifest Destiny. During that time, cattle would progress from nonexistent in the northwestern territory to being intensively produced across the expanding U.S. territory.

In *Cattle Country*, therefore, I study the themes connected to the land and animals, what is often described as "the environment," as well as the humans on that land, examining how race became applied in ways that would privilege one form of agriculture over others. The theoretical approaches of critical race studies, ecocriticism, and food studies—including agriculture and animal studies—intersect in narratives about cattle and the expanding U.S. border as representations of the nation's settler colonialism. Paul Outka describes the natural environment and race as "perhaps the two most perniciously reified constructions in American culture."[11] These constructions intersect in terms of nineteenth-century U.S. agri-expansion. The modern incarnations of critical race studies and environmental criticism occurred in the 1960s, but the need for people to write about their natural environment and human interactions within that environment predates modern history. Historical forms of racial categorization created a structure that privileged Euro-American cultural practices over others as seen in the works of people of color. In *Cattle Country* I focus on narratives describing Native American characters, as well as Mexican Americans, African Americans, and Asian Americans, as they negotiate contested spaces throughout U.S. agri-expansion. In addition, the land that would come to be known as "the natural environment" was constructed in the nineteenth century in particular through the creation of national parks, starting with the inauguration of Yellowstone National Park in 1872 and solidified by the formation of the National Park Service in 1916. These protected spaces developed in such a way that indigenous communities were no longer allowed access to their own forms of agriculture, in the interest of protecting a Euro-American concept of "natural space." The "land" and the "environment" became divided into cultivated and natural, agricultural and wasted or protected. Throughout this period, people of color and environmentalists worked to fight against forces seeking to subjugate both groups under one dominant society. Michael Omi and Howard Winant and LeAnn Howe, among others, analyze the application of race and power throughout U.S. culture, including culinary culture.[12] The authors I discuss in *Cattle Country* engage with repeated Native American removals in terms of the development of U.S. agriculture, as well as the post-frontier racial and class-based U.S. cuisine reliant on cattle. Furthermore, I apply the ecocritical work of Lawrence Buell, Paula Gunn Allen, and Annette Kolodny to argue that the authors discussed in *Cattle Country* helped create a developing awareness of and concern for resource depletion, habitat destruction, and a wasteful overproduction of a particular monoculture of food that that would eventually lead to the food crisis of the twenty-first century.[13]

Food studies has developed into a dynamic method of communicating culture in shared stories and histories. "Food history is a theme of world history," argues Felipe Fernández-Armesto, "inseparable from all the other interactions of human beings with one another and with the rest of nature."[14] The stories we tell and the foods we prepare are two foundational components of culture. Identifying the narrative elements in food and literature, Carolyn Korsmeyer recognizes the importance of taste. Both forms of culture take a set amount of time to process, for example; there are regulated steps and recipes in both, and both can be seen as art forms.[15] In either case, a narrative is being communicated, the senses are engaged and developed, and people share in the experience. Identity and nationality are defined by culture: literature, art, music, politics, religion, and, importantly, cuisine and agriculture. Margaret Mead's work on changing U.S. food habits in the 1940s—the facts she offered about the interdisciplinary ways that culture grows because of these changes—is equally applicable to the nineteenth century.[16] As writers worked to standardize certain cultural traits and develop new ones during this time, they engaged in a discourse of culinary and agricultural practices, as Susan Fenimore Cooper and Henry David Thoreau would describe. In the nineteenth century the United States was working to define itself through its literature as well as agricultural practices, as part of its cultural movement of self-definition, and I place myself at this intersection of literature and food studies.

Throughout *Cattle Country*, then, I study the use of cattle by these nineteenth-century authors. *Cattle*, of course, is a broad term, referencing a number of livestock and wild animals. I use the term to refer to particular bovines: meat and dairy cattle, oxen, and bison.[17] I frequently juxtapose domestic cattle against their wild equivalent, the bison. Barbara Nelson describes the lineage of cattle leading up to the modern breeds. "The modern cow descended from extinct wild ungulates like the African, Asian, and European aurochs (*Bos Taurus primigenius*). Scotland's legendary, long extinct, wild white cattle were also ancestors, as are India's endangered beautiful red Gaur (*Bos gaurus*). The cow's family tree includes Caesar's urus, Indonesia's banteng, and the hairy wild yak (*Bos grunniens*). Even the European and Asian wisent (*Bison bonasus*), a small, light-colored buffalo, contributed to the wild gene pool."[18] Donald Sumner notes that cattle (including cows, oxen, and bison) are significant agriculturally at the global and regional levels for meat, milk, and labor: "cattle tend to be where people are."[19] Cattle have always been global, perhaps even expansionist. They need space for grazing, mimicking U.S. expansion, unlike other domestic animals like pigs or chickens. Cattle were also valuable to families moving west along the U.S. frontier, providing food through meat and dairy products and labor in the form of oxen, as seen in the painting *American Progress*. In fact, John L. O'Sullivan coined the term Manifest Destiny in 1845 in part to help increase the territory available for U.S. farmers—to promote the expansion of domestic livestock. This movement intensified after the Civil War, urbanization, industrialization, and advances in transportation. Eventually, these agricultural revolutions led to the meat-processing plants of midwestern cities made famous in works like Upton Sinclair's *The Jungle* (1906), almost exactly a century after the Lewis and Clark Expedition opened up the possibility of westward expansion to the nation. Bison, on the other hand, are a stark and poetic contrast to domestic cattle throughout the nineteenth century. By the end of the century, bison would earn protected status at Yellowstone National Park as part of the national park movement, becoming a romanticized symbol of the nation in their own way after having been nearly driven to extinction to make room for the overly domesticated cattle.[20]

If cattle are the representative nineteenth-century U.S. animals, the U.S. farmer—nearly always seen as Euro-American in these narratives—is the iconic laborer. The inception of the nation and its earliest rationale for expansion promoted a kind of Jeffersonian agrarianism, the concept of a society of small-scale farmers who worked to produce most of their own food and other goods, selling or bartering any surplus. In *Notes on the State of Virginia* (1787), Thomas Jefferson famously declares: "Those who labour in the earth are the chosen people of God, if ever he had a chosen people, whose breasts he has made his peculiar deposit for substantial and genuine virtue. It is the focus in which he keeps alive that sacred fire, which otherwise might escape from the face of the earth."[21] Indeed, he continues, other forms of industry would lead to the "corruption of morals."[22] Therefore, when Jefferson decided to purchase territory from France, his idea was to create more arable land to be used potentially as part of his project of land stewardship.[23] This stewardship, though, included the enslavement of African Americans for their labor, and this internal movement of enslaved peoples became part of westward expansion, seen in the annexation of Texas and the Mexican-American War.[24] Meanwhile, Native American forms of agriculture, while varied, were almost always considered to be oversimplified examples of "hunting and gathering" rather than as acceptable forms of agriculture, including their animal husbandry practices. This was made clear through a variety of nineteenth-century policies including Tecumseh's defeat in 1813 during James Madison's presidency, Native American removals starting in 1830 under Andrew Jackson, and the efforts during Ulysses S. Grant's tenure to, as stated by Philip Sheridan in 1876, "destroy the commissary" of Native Americans of the Great Plains region by removing the bison population that they depended on.[25] The rationale was that indigenous communities, because of their use of a less favored form of animal husbandry, were less deserving of the vast expanses of land making up North America. The United States took advantage of a privileged form of Euro-American agriculture—conveniently forgetting indigenous forms of agriculture that had preserved them in the early years of colonization—as an excuse to expand into the western lands of the continent.

Any work discussing indigenous communities and the environment has to take into account the uneasy history in U.S. culture considering the idea of Native Americans or American Indians as being more connected to the nebulous concept of "nature" than Euro-Americans. Concepts like race and the environment have particularly loaded meanings when applied to Native communities following settler colonialism. Even when depictions are intended seemingly positively, this is a thorny issue. I engage with several nations in *Cattle Country*, including the Wampanoag,

Iroquois, Pawnee, Penobscot, Paiute, and Kumeyaay. These communities had widely different styles of agriculture depending on their region: growing corn, beans, and other crops, harvesting pine nuts and roots, fishing and harvesting shellfish, as well as hunting deer and bison. They represent different ways to engage with lived environments to produce and prepare food, clothing, and other commodities, ways that developed sometimes over thousands of years. Charles C. Mann discusses, for example, how the Cahokia of the Mississippian peoples created one of the largest communities in North America, the largest north of the Rio Grande, near modern-day St. Louis, Missouri, based on farming.[26] While this agricultural society rose and collapsed before Euro-American contact, other indigenous agricultural methods have since proved to be more appropriate in lasting ways than the industrialized agriculture promoted by the United States since the nineteenth century.[27] Largely, though, an all-encompassing idea of the "noble savage" became popular during the nineteenth century. Some examples of Native characters in literature—as in James Fenimore Cooper's *Last of the Mohicans* (1826) or *The Pioneers* (1823)—did much to popularize this stereotype: people at home more in the wilderness than in civilization and doomed to vanish because of an inability to adapt to modernization. This trope has been used frequently in U.S. literature and culture.[28] Indeed, I maintain that Sarah Winnemucca Hopkins herself manipulated stereotypes of the "Indian Princess" to her benefit while working to improve the living conditions of her people, the Paiutes. In *Cattle Country*, I mostly discuss the impact of agriculture and expansion on Native American communities—as well as on Mexican American, African American, and Asian American communities—as all are ultimately compared to the Euro-American farmers and their representative cattle.

Ultimately, I argue that my selected authors showed cattle to be the icon of the United States, and that such a decision was of the deepest significance to U.S. farmers and the nation. Animals are part of the environment, and how humans work with them for food, clothing, and other material goods is one of the most significant ways to show their impact. Steven Stoll somewhat tongue-in-cheek calls U.S. farmers "parasites of soil" moving ever westward for "fresh blood."[29] The idea of "the West" being connected to "blood" is apt. U.S. agriculture has long been extractive, bleeding territories dry before moving on to new ones. In addition to environmental concerns, such agri-expansion has been an anxiety in terms of food justice. Janet Fiskio reads food justice as a twentieth-century movement, though many of its concerns and critiques of Euro-American agrarianism apply as well to the nineteenth century. In the following chapters, writers and characters must negotiate events and policies, including the failure of Reconstruction and the Dawes General Allotment Act, that affected U.S. agriculture. Policy makers used livestock husbandry and other forms of agri-expansion to justify the forcible removal of Native American communities from their lands. Communities—indigenous and immigrant—ended up working in an "agrarianism of the margins" in a way that required them, as well as the authors engaging with U.S. agriculture, to navigate this agri-expansion.[30] Fiskio warns against privileging a single idea of agriculture, saying rather that there are multiple successful ways to grow and produce food, including animal husbandry. The question of appropriate land stewardship appears in some form in all the works discussed in *Cattle Country*. I narrow my focus to the intersection of cattle, the U.S. farmer, the people of color who inhabit the land, and the land itself. Throughout, I consider works of fiction, novels, and short stories in addition to journals, travel narratives, nature writings, and images. Having introduced the issue of expansion and agriculture in the 1806 journals of the Lewis and Clark Expedition, I study a span of roughly one hundred years, concluding with Sinclair's 1906 publication of *The Jungle* and Winnifred Eaton's *Tama* (1910) and *Cattle* (1924). I analyze some of the most popular authors of the time along with some of the century's most enduring cultural critics. Throughout, various and ubiquitous appearances of cattle in the literature of the nineteenth century show them to be an indispensable cultural symbol.

In *A Tour on the Prairies* (1835), Washington Irving discusses the Far West of his time, the Indian Territory that would become eastern Oklahoma, and he continues to write about the developing U.S. West in *Astoria* (1836) and *The Adventures of Captain Bonneville* (1837). Upon his return from a seventeen-year European sojourn, Irving felt pressured to prove himself truly "American" once again by writing of his visit to lands populated by the Pawnee nation after their removal west of the Mississippi River. Irving's idea of "native" lands is therefore thrown into relief. Moreover, Irving wrote at the beginning of the frontier period, which grew in importance throughout the nineteenth century, until U.S. historian Frederick Jackson Turner famously reported of its "closing" in 1893. Irving visited the Great Plains at a transitional time, therefore, between the reign of the bison and that of domestic cattle. His goal upon the visit to the territory was to hunt bison, commonly referred to as buffalo. Toward the end of *A Tour on the Prairies*, though, Irving directly suggests converting the Great Plains region from home to free-ranging bison to domestic cattle living on small farms like those of the East. He goes on to describe the advantages of cattle in the West in key moments of his other western writings. In each case, Irving recommends applying U.S. forms of agriculture to the increasing nation, and his great popularity meant that his romanticized comparisons of bison and cattle and the physical increase of Jeffersonian agrarianism would attract a wide audience, especially the new generation of settlers that followed him west. In chapter 1, "Washington Irving, Cattle, and Indian Territory," I argue that Irving encouraged nineteenth-century agri-expansion through his depictions of cattle—and their greater suitability to the Euro-American farmer's concept of land use than bison proved to be—in his three western narratives.

As Irving was writing about the developing U.S. West, James and Susan Fenimore Cooper, father and daughter, were also writing about the West and the Otsego region of New York during the first half of the nineteenth century. In his Leather-stocking novels and her nature writings, they describe an idealized village life, at the same time warning against resource depletion and the degradation of the environment, over a century before such terms became common. Like Irving, they also wrote at a transitional moment, as the United States was moving from Jeffersonian agrarianism to Jacksonian expansionism. I study their conflicting attitudes toward the growing nation in chapter 2, "Civilizing Cattle in the Writings of James and Susan Fenimore Cooper." In their work, the Coopers privileged agrarianism. While Leather-stocking might have lit out to the territories at the end of The Pioneers (1823), with readers finding him again in the Great Plains region in The Prairie (1827), James Fenimore Cooper decidedly does not recommend that move to the rest of the country. The emigrating Euro-Americans in The Prairie, in fact, return east by the end of that novel. Moreover, while Cooper relies on stereotypes of Native American characters, he does provide diverse perspectives of agricultures and ways of living other than those of the Bushes and their companions in The Prairie. His daughter Susan Fenimore Cooper would return to describing the Cooperstown region in Rural Hours (1850), where she argues against the move toward deforestation and other forms of waste she saw beginning in the West. She also describes a family visit to a neighboring farm—cattle looming large in this extended scene—recognizable as her ideal form of eastern U.S. agriculture. Together, the Coopers argue for conservation and stewardship as a necessary part of U.S. agriculture from Cooperstown, New York, to the increasing U.S. territories of the West.

As a contrast, Henry David Thoreau does not write primarily about cattle or the West in his late full-length nature writings, which focus rather on moose and clams in the Northeast. In chapter 3, "Henry David Thoreau, Regional Cuisine, and Cattle," I interrogate two of Thoreau's narratives, *The Maine Woods* (1864) and *Cape Cod* (1865), as they privilege regional foods and animals. Indeed, the idea of eating something regionally specific as opposed to the representative cattle was a foundational element of Thoreau's purpose for his journeys—as important to him in the Maine woods and Cape Cod as his beans and the woodchuck were in *Walden* (1854). In his travel narratives, including *A Week on the Concord and Merrimack Rivers* (1849), Thoreau focuses on deep awareness of specific bioregions through the production and consumption of New England's "grossest groceries" at a time when the nation was at one of its most dramatic moments of expansion, following the nature writings of Susan Fenimore Cooper. As he cooks his own harvested clam or makes a stew of boiled tongue from a moose that his party had hunted, Thoreau questions and helps to define an early U.S. gastronomy at a time when the nation was in one of its most intense moments of self-definition. Meanwhile, he researches the natural history of the regions he visits, including their agricultural history, learning from Penobscot guides as much as from his own reading. As Ralph Waldo Emerson would call for a uniquely U.S. literature in "The Poet" (1844), Thoreau develops his own appeal for a corresponding U.S. cuisine; however, his focus is on regional forms of agriculture, not a cattle-based monoculture. Thoreau considers an essential part of the venerability of his excursions to be the discovery and sampling of indigenous foods that have a special relationship to place, that weave the eater into the local environment, and that honor Native traditions.

Later in the nineteenth century, Sarah Winnemucca Hopkins would use cattle to describe her experience of U.S. settlement of the West as an example of how one indigenous community had to dramatically alter itself because of the encroachment of U.S. settlers and their cattle. These struggles permeate her most famous work, *Life among the Piutes: Their Wrongs and Claims* (1883). In addition, some of her most passionate rhetoric appears in newspaper articles published throughout her career, as well as in her 1884 congressional testimony. The traditional food-ways of the Paiutes (Winnemucca spells it Piutes and Pah-Utes in her titles) were diminished through expansionist U.S. land policies, a more intensified version of the policies criticized in Thoreau's writings. The Paiutes were driven to a commodity economy, forced to learn U.S. systems of agriculture that included cattle ranching, and eventually driven to starvation on a series of reservations at the hands of corrupt land agents. In chapter 4, "Cattle and Sovereignty in the Work of Sarah Winnemucca Hopkins," I argue that Winnemucca describes the settlers' actions against her people in terms of cattle in two key ways. First, cattle became competition for resources: the Paiutes would lose their traditional lands to make way for cows and other livestock which then changed the environment through grazing. In addition, cattle became comparative figures, as Winnemucca's commu-nity itself would come to be treated as merely another form of stock. Winnemucca uses cattle deliberately in her tribalography, essays, and testimony, then, to gain sympathy from white audi-ences and U.S. government officials in order to help her people, the Paiutes, during a period of especially brutal treatment. She works throughout her adult life to try to restore some form of food sovereignty to her people.

In her novel *The Squatter and the Don* (1885), María Amparo Ruiz de Burton extends Winnemucca's expansionist concerns to the Californio/a culture leading up to and following the 1848

Treaty of Guadalupe Hidalgo, which ceded the lands including California to the United States from Mexico. In chapter 5, "The Cowboys *Are* Indians in *The Squatter and the Don*," I maintain that cattle had become a dominant symbol in terms of land and food sovereignty in the late nineteenth-century West, specifically the Mexican-American region of California. Ruiz de Burton's primary conflict pits her hero, Don Mariano Alamar, against the U.S. "squatters" and connects back to cattle; her "vaquero" or "cowboy" terminology is as integral in the novel as her use of legal documents. Almost every scene involving cattle—outdoors and engaging with nature—also involves Native American characters. The indigenous peoples of the San Diego County/northern Baja California region are the Kumeyaay (Diegueño). They perform much of the technical and skillful work with cattle throughout the novel. Ruiz de Burton has been criticized for her treatment of Native American characters in her novels. In *The Squatter and the Don*, though, she shows their great skill with cattle, as well as the obvious importance given to cattle as part of the economy and culture of California during this time. Therefore, while she does not give her Kumeyaay the time or attention she affords her Euro-American figures, she cannot help but value these skilled vaqueros existing on the periphery of her novel. Ruiz de Burton both criticizes U.S. policy and reclaims the vaquero as the origin of the cowboy—one of the nation's most mythologized figures—through her novel's Californio/a and Kumeyaay characters as they work with cattle.

U.S. cattle monoculture through westward expansion also had devastating effects for the South. Charles Chesnutt's stories collected in *The Conjure Woman* (1899) and his other *Conjure Stories* complicate the nation's post-Reconstruction food culture and the dominance of cattle by examining agriculture in the South at the end of the century, a region defined by its connection to pigs and chickens rather than to cattle. Indeed, Booker T.

Washington would promote beef as the ideal dining fare at the Tuskegee Institute in an aspirational gesture by the end of the nineteenth century, working to distance African American culture from the stereotypical cuisine represented by pigs and chickens. Livestock were key resources for early colonists, with pigs and chickens often being more important food animals than cattle before the nineteenth century. Cattle would really only develop into the nation's chosen food animal with the territorial expansion to the West that included the issue of the extension of slavery. This focus on the West, though, to some extent displaced the South; as cattle came to dominate, regions inappropriate for grazing were relegated to the margins. Chesnutt links people and livestock throughout his conjure stories set in North Carolina. Interestingly, though, he more often describes scenes with pigs and chickens than with cattle in terms of cuisine. Chesnutt responds to another issue of cattle-based U.S. expansion that neither Winnemucca nor Ruiz de Burton discusses. Elements of southern racial histories were intimately connected to their food and labor: the South became connected to pigs and chickens in a similar way that the larger nation was becoming associated with cattle. Characters in Chesnutt's *Conjure Stories*—free and enslaved, men and women—are given and denied power and resources through their engagement with particular food animals. In chapter 6, "Southern Cuisine without Cattle in Charles Chesnutt's *Conjure Stories*," I analyze the ways that Chesnutt weaves together elements of race, land, and U.S. animal husbandry as his stories question what is for dinner in the post-Reconstruction U.S. South.

I bring my project into the twentieth century and to a transnational scale in chapter 7, "Industrial-Global Cattle in Upton Sinclair and Winnifred Eaton." The obvious extension of my discussion on the intensification of animal husbandry in nineteenth-century U.S. literature, Sinclair's *The Jungle* (1906) details the issue of mass production of meat, largely beef, in the slaughterhouses of Chicago and other midwestern cities near the turn of the century. This grisly work shows the culmination of nineteenth-century expansion of the cattle and beef industry. While the Beef Trust gains immense wealth for a few, a vast number of Chicago's most vulnerable workers—immigrants and people of color—suffer immensely through the industrialization of agriculture. In addition, the environment is threatened immeasurably through this form of agriculture practiced in the stockyards of Chicago. The impact of the novel has been nearly unparalleled in U.S. literature and history and is the foundation for several of the nation's food safety regulations. In addition to this work of Progressive Era muckraking, I also study global narratives of beef in the novels of the Chinese Canadian author Winnifred Eaton Babcock Reeve, who navigated some of the most dramatic U.S. anti-immigration attitudes following the Chinese Exclusion Act of 1882 by fashioning an identity for herself as a Japanese author. Eaton moves from essentializing beef as necessary Euro-American male fare in works like *Tama* (1910) to her own work as owner of an Alberta cattle ranch, described in her novel *Cattle* (1924). Cattle production also was growing globally in the first decades of the twentieth century. Canada's Great Plains are geographically similar to those of the United States, and it became a simple practice for Canada's cattle ranchers to traverse national borders. Indeed, her novel's villain appears to be a cross between one of Winnemucca's dispossessing cattle ranchers and Sinclair's Beef Trust.[31] By the first decades of the twentieth century, these two authors were showing the importance of cattle husbandry outside of the geographic borders of the United States.

It is obvious in the twenty-first century how one particular form of agri-expansion that developed in the United States over the long nineteenth century has proven to be dangerous for the planet in terms of environmental degradation, the health of human and nonhuman actors, and class and racial concerns across the globe. It is important, therefore, to study how cattle husbandry was being envisioned in the nineteenth century in order to contextualize its results, especially concerning race and ecology. Several U.S. writers of the time applauded the idea of agri-expansion in the nineteenth century, though not without reservations. Some, rather, were as critical of what they saw as dangerous stewardship practices as anyone in our current moment. Additionally, certain authors described agri-expansion as they witnessed it—an inevitability in which they tried to survive and, if possible, thrive. *Cattle Country* provides a narrative history to the agri-expansion in the United States, specifically of the iconic cattle, during this critical century. There has been surprisingly little written on early U.S. attitudes about this representative animal; therefore, I show not only what cattle would come to mean to U.S. farmers but also how they were instrumental in developing particular ideas of race and identity throughout nineteenth-century narratives.

ONE

Washington Irving, Cattle, and Indian Territory

In one of Washington Irving's best-known stories, "The Legend of Sleepy Hollow," from *The Sketch Book of Geoffrey Crayon, Gent.* (1819–20), the attitudes of the character Ichabod Crane foreshadow what would become Irving's own—as well as the nation's—thoughts about western expansion in the nineteenth century. In the story, Crane desires food and material wealth, privileging agriculture and animal husbandry of a kind represented by the prosperous farmer, Baltus Van Tassel. Cattle and horses are the animals most representative of the U.S. West during the nineteenth century, and when Crane attempts to ingratiate himself to the villagers, he does so in a manner that includes working with these animals: "He assisted the farmers occasionally in the lighter labours of their farms, helped to make hay, mended the fences, took the horses to water, drove the cows from pasture, and cut wood for the winter fire."[1] Notably, the cattle are connected to their need for and use of pasturelands, large enough grassy spaces that multiple people are required to help move the cattle across them. Later, in a scene that predates the most dramatic period of westward expansion, Irving draws a link between the acquisitive character of Crane and the national expansion of a privileged, Euro-American kind of agriculture, what I call agri-expansion: "Nay, his busy fancy already realized his hopes, and presented to him the blooming Katrina, with a whole family of children, mounted on the top of a waggon loaded with household trumpery, with pots and kettles dangling beneath; and he beheld himself bestriding a pacing mare, with a colt at her heels, setting out for Kentucky, Tennessee, or the Lord knows where!"[2] Crane imagines for himself the expansionist tendencies of the nation, and he describes a frontier life with a family, a wagon, and domestic animals in tow. In this sketch, Crane—and perhaps by extension Irving—is already looking west, and the nation would soon follow suit. The expansion of Euro-American animal husbandry, along with the removal of any competition for those resources, was a significant issue in the northeastern United States even before the drive to move west. In another piece from *The Sketch Book*, "Philip of Pokanoket: An Indian Memoir," Irving gives a sympathetic portrayal of Metacomet (spelled Metamocet in the story), called King Philip by the colonists, the Wampanoag leader of King Philip's War.[3] In this sketch, Irving calls the forests around Pokanoket "impervious to anything but a wild beast, or an Indian" and "impracticable to the white man, though the Indian could thrid their labyrinths with the agility of a deer."[4] Deer and other beasts are seen to be freely living in these forests throughout the conflict, as are the Wampanoags, and these "wild" humans and animals are contrasted to the Euro-Americans and their domestic cattle. In this reading, then, one form of agriculture and resource management is favored over another.

The privileging of one form of agriculture over any other is important to the narrative created around westward expansion in the first half of the nineteenth century. In the idealized village of Sleepy Hollow, perhaps similarly to the village of Pokanoket at the time of King Philip's War, the representative farmer, Van Tassel, does not seek to extend his bounty beyond his own borders; however, Crane is shown to be as much of a devouring "anaconda" in this scene as at any point in "The Legend of Sleepy Hollow."[5] The anaconda-like Crane, whom Daniel Hoffman calls

"Ichabod Boone," tends toward the concept of wealth associated with expansion, of exploiting new lands along the frontier for his own gain.[6] In this comparison, it is Metacomet that seeks to maintain his borders from the U.S. expansionists in a more dramatic fashion. As Irving writes of domestic cattle replacing wild food animals, he also describes the replacing of Native communities with Euro-American ones, and his popularity as a writer would make his language particularly potent during this early stage of agri-expansion. Mark Burns observes that even as Irving challenges misrepresentations of Native Americans elsewhere in literature, he helped to create that stereotypical role in his popular sketches, "Philip of Pokanoket" and "Traits of Indian Character."[7] Following the success of *The Sketch Book* and his return to the United States, Irving would go on to write popular narratives of the West in which he continued to describe an agri-expansionist transition from the traditional methods of planting and harvesting used by various Native American nations and tribes to a more intensive, and eventually mechanized and industrialized, system of animal husbandry. Irving describes the exploitation of new lands across the North American frontier, the replacing of wild animals with domestic ones involved in that move, and the removal of Native American peoples for Euro-American settlers outlined in *The Sketch Book* more directly in his western writings *A Tour on the Prairies*, a novella published in *Crayon Miscellany* (1835), *Astoria* (1836), and *The Adventures of Captain Bonneville* (1837). I argue that Irving's writing project, as part of the larger cultural move-ment of the first half of the nineteenth century, was to help define a unique U.S. culture through descriptions of specifically cattle-based animal husbandry in the West and romanticized depictions of Native American characters and cultures, even as he pronounced them to be vanishing. In so doing, he helped further the national project by promoting westward expansion geared toward agriculture—a project that would be developed and intensified throughout the century.

Important for both Metacomet and, later, for Irving in his own western travels is the contrast between the wild animals and domestic livestock in these regions. Wildness is represented by deer in his sketches and the American bison in his western narratives; meanwhile, Irving's most important domestic animals are cattle, specifically, as well as horses throughout his writings. In this way, Irving was following historical patterns. When the Europeans arrived in the Americas they brought their foods—including domestic animals like cattle, pigs, and chickens—with them. These foods added to and sometimes conflicted with the wealth of native foods discovered through the Columbian exchange. In many ways, livestock became the cause for struggle between the agricultural methods of Euro-Americans and Native Americans. Euro-Americans consistently described Native Americans as hunter-gatherers, though they had a far richer engagement with the land than this phrase implies. Indeed, many nations grew the iconic "Three Sisters" crops of corn, beans, and squash; meanwhile, livestock played a significant role in early conflicts between colonists and indigenous communities such as King Philip's War of 1675. In this war, for example, in which both the Native American communities and colonists were killed by the thousands, the livestock that was so important to the Euro-Americans were also killed en masse. Livestock became a dramatic form of collateral damage during this and other early North American conflicts. They represented Euro-American settler colonialism, literally and symbolically replacing the wild game of the region. During this early period of North American history, therefore, livestock had already come to represent Euro-American colonialism, and the importance of cattle to westward expansion would only grow over the nineteenth century.[8]

Throughout his U.S. writings, then, Irving would combine his vision of Leo Marx's pastoralism with Frederick Jackson Turner's frontier thesis. Irving's U.S. writings correspond to Marx's "American fables" in terms of Thomas Jefferson's pastoral ideal. However, Irving writes during the period of Andrew Jackson's expansionism, and that influence is present in his work as well. The frontier would grow in importance throughout the nineteenth century, until Turner would famously write of its "closing" in 1893. Conceptualizing his famous thesis, Turner describes cattle as being ideal—almost necessary—for agri-expansion. The iconic image of the cowboy on the range, therefore, was an indispensable part of the development of the U.S. frontier, as would become most apparent during the cattle boom of the late nineteenth century. Turner places cattle, animals that he had already referred to as part of the iconic image of the U.S. West, along with the cowboy and the open range, near the apex of his civilized concept of the frontier: "Watch the procession of civilization, marching single file—the buffalo, following the trail to the salt springs, the Indian, the fur trader and hunter, the cattle raiser, the pioneer farmer."[9] For Turner, bison and domestic cattle occupy nearly opposite poles in his idea of frontier civilization, and Irving would describe a similar belief throughout his U.S.-based writings. Those Native American nations and tribes living in the frontier territories of the Great Plains were seen to exist somewhere between the two. Similar to Turner's conceptualization of agriculture as part of civilization's progress, Irving's descriptions of "Indian character" emphasize his concerns about agriculture and the use of resources as they developed more fully in his later western narratives.[10]

In "Traits of Indian Character," therefore, Irving falls into the language popular during his time of describing northeastern Native American communities as both "noble savages" and inevitable victims of Euro-American civilization. Irving romanticizes the "North American savage" as "sublime" along with the "vast lakes, boundless forests, majestic rivers and trackless plains."[11] The final item in Irving's list is the seemingly "trackless plains" that would be the subject of his western narratives. The northeastern peoples he discusses in this essay are equated with nature, contrasted to the encroaching Euro-American "civilization," which he describes as the opposite of nature. This binary is reaffirmed as Irving, in trying to compliment idealized Native American figures, writes, "The colonist has often treated them like beasts of the forest; and the author has endeavoured to justify him in his outrages. The former found it easier to exterminate than to civilize; the latter to vilify than to discriminate."[12] Irving makes two symbolic moves in this statement: he notes the tendency in narrative to compare Native Americans to "beasts," and he argues that the proper treatment for Native Americans—the more humanitarian path—is to "civilize," or assimilate them. However, his implication is that if these northeastern nations fail to assimilate to the Euro-American style of agriculture, to be consumed into the growing nation, they would inevitably be destroyed. In so doing, as an author, though ironically not the author he criticizes in this passage, Irving foreshadows Jackson's Indian removal plans of the 1830s.[13]

Returning to "Philip of Pokanoket," then, Irving continues his theme of land use, animals, and humans, this time during King Philip's War between the Wampanoags and Euro-American settlers. "The report of a distant gun would perhaps be heard from the solitary woodland," he writes, "where there was known to be no white man; the cattle which had been wandering in the woods, would sometimes return home wounded; or an Indian or two would be seen lurking about the skirts of the forests, and suddenly disappearing."[14] Three figures are compared in this passage: Euro-Americans, Wampanoags, and cattle. Their setting is the woods around Pokanoket—a territory not yet exposed to U.S. animal husbandry. The Wampanoags are most successful at navigating the forest; they are read as most "wild" and most adapted to wilderness spaces. The Euro-Americans are the reverse, entirely unsuccessful in any attempt to manage the region; they are not to be seen. Indeed, they are represented as feeling under threat. Cattle, though, are somewhere in between these human populations. They attempt to forage in the forest but risk becoming "wounded" when they try to do so rather than grazing in their more familiar pastures. Irving makes an almost ecocritical argument by emphasizing nature and the power of the natural world in this scene, but his ultimate conclusions remain in favor of Euro-American civilization in the form of one particular kind of agriculture, one that includes the domestication of animals, and the assimilation of the Native American tribes and nations to that particular form of agriculture.

During the first half of the nineteenth century the United States worked to establish a distinct national identity, and the indigenous communities already living in North America became significant parts of that identity. Native Americans thus were sometimes described in an elevated neoclassical manner, their virtues read in comparison to earlier European civilizations as part of the nation's move to cultural nationalism following the War of 1812. At that time, Native Americans were represented in U.S. literature under the myth of the "noble savage," not unlike a Roman or Spartan figure. Irving was part of this cultural movement, along with James Fenimore Cooper, another of its greatest practitioners. Interestingly, Irving's romanticization changed following Jackson's Indian removal policies, which showed that Native Americans had not in fact vanished at all. Throughout both eras, Irving provided often sympathetic descriptions of his Native characters, especially when compared to other early nineteenth-century writ-

Fig. 2. Charles Turner produced this portrait of Washington Irving ca. 1825 while Irving was in Europe, working to measure up to his ideal of British gentlemanliness, while also remaining a staunch U.S. patriot. Smithsonian American Art Museum, gift of Olin Dows.

ers. He modified his descriptions of certain Native Americans in his western travel narratives, though, after 1830; characters in these works became either "hostile" (Sioux, Blackfeet, Crows) or "friendly" (Walla Wallas, Nez Perces, and Snakes [Shoshones]). They were no longer universally described as "noble." In his writings of Native Americans set against the beginnings of U.S. agri-expansion, Irving moved from romanticization to a more complicated form of engagement.[15]

When Irving published his Native American sketches in *The Sketch Book*, he was living in Britain. At the same time, Irving—the youngest of eight children—was a decided patriot. He always remembered meeting his namesake, George Washington, when he, Irving, was six years old, and throughout his writing career, Irving united his U.S. patriotism with a determination to measure up to British ideals of gentlemanly behavior. His pen name, Geoffrey Crayon, Gentleman, is an example of this amalgamation. A portrait made of Irving years before his western travel shows Irving as a man working to be the "gentleman" of both his pen name and namesake (see fig. 2). As part of that desire to measure up, perhaps, Irving followed and promulgated tropes of his time. Irving's work in *The Sketch Book* was part of the overall development of the U.S. nation and culture. Irving's writings are place-based, and he became one of the first celebrity U.S. writers in large part because of his ability to communicate a powerful sense of region in his work. Irving was writing at a pivotal period for U.S.–Native American relations. Between 1815 and the 1830s, national attitudes toward Native Americans along the frontier changed from the desire for assimilation and romanticization to that of removal. Situated during this period, Irving's writings promoted western expansion, cattle raising, and therefore implicitly Jackson's Indian removal, upon his return to the United States and the publication of *A Tour on the Prairies*, *Astoria*, and *The Adventures of Captain Bonneville*.[16]

"Little Stretch of Fancy":
Frontier Farms in *A Tour on the Prairies*

In *A Tour on the Prairies* (1835), Irving discusses the far west of 1832, the Indian Territory in what would become eastern Oklahoma. Upon his return from a sojourn of seventeen years in Europe, during which time he had been involved in the successes and failures of the family business, Irving felt pressured to prove himself a true "American" once again. He wanted therefore to visit the Indian Territory before the multiple removals ultimately resulted in disaster to the communities currently living there. Irving worried that "those great Indian tribes" were "now about to disappear as independent nations, or to be amalgamated under some new form of government."[17] He documented numerous interactions between cultures along the journey—contrasting these sketches to supposed violent encounters that, in fact, never took place. Irving satirizes the early nineteenth-century trope of the Native American adventure tale. In addition, he puns on the idea of a "native country" in his preface.[18] He has newly returned to the land of his birth following a long absence. However, he is traveling in recently appointed Indian Territory, throwing into dramatic relief the question of "native" lands. Irving plays on the term "native lands" repeatedly throughout his work, and this theme can be seen in comparisons between Native Americans and Euro-Americans, as well as animal examples of bison and cows. While Irving writes for the most part considerately regarding Native characters, he ultimately chooses the agri-expansionists as his narrative heroes. His tour occurred at a time when the nation was challenging territorial and economic borders, looking west toward an imagined frontier. His work studies these early stages of westward agri-expansion, but it also provides an incisive look at some of its consequences. In *A Tour on the Prairies* specifically, Irving imagines replacing the bison with domestic cattle as a way to offer an impetus for western settlement and the physical expansion of U.S. borders.[19]

Irving describes his travels to the Indian Territory on a month-long hunt for bison with friends. One of these friends was Henry Leavitt Ellsworth, employed by the Jacksonian administration as a commissioner of Indian tribes following the 1830 Indian Removal Act. At this time, Jacksonian expansion was becoming more established in the nation, and new territories were seen to be desirable for agricultural and economic development. This meant that the territory west of the Mississippi was undergoing an extreme transformation. In fact, Ellsworth would go on to establish the Department of Agriculture. The Indian Territory seemed "designed expressly for agriculture" to U.S. settlers, according to Waverly Root and Richard de Rochemont, who add, "farming meant cattle raising, an important activity in the territory even before the beef producers had a legal right to be there."[20] Popular narratives that promoted this nineteenth-century idea of westward expansion—especially by the famous author of "The Legend of Sleepy Hollow"—provided powerful motivation to U.S. settlers. Irving describes his party's journey from its base at Fort Gibson on October 10 through its return on November 8, 1832.[21] The period in which Irving traveled to the Great Plains, as well, was a time when that region was becoming romanticized in the East. Well-known authors like James Fenimore Cooper and William Cullen Bryant had already written about the region in *The Prairie* (1827) and "The Prairies" (1832), respectively.[22] In "The Prairies," for example, Bryant describes bison as domesticated as much as cows had been for the "mound builders," or the Mississippians. For him, the prairies "Nourished their harvests, here their herds were fed, / When haply by their stalls the bison lowed, / And bowed his manèd shoulder to the yoke."[23] Irving follows Cooper and Bryant in his thrilling and ironic descriptions of wildness, as opposed to civilization, and Native American nations and tribes on the Great Plains. Ultimately, though, Irving was disappointed in each of these goals. In *A Tour on the Prairies* he comes to realize that the Native Americans he encounters are not dangerous, but rather often friendly. The wildness he seeks is not pristine, nor is it hospitable. In addition, bison are replaced with domestic cows in his imagined future for the region. While the ultimate bison depletion was carried out later in the century, as a form of warfare against the Great Plains nations and tribes dependent on them as a necessary food source, the idea of expansion—of U.S. Americans with cattle ranches ultimately replacing free-roaming bison and the humans dependent upon them—originates far earlier with works like Irving's.[24]

Almost the first thing Irving does in *A Tour on the Prairies* is to call the Indian Territory "a vast tract of uninhabited country."[25] At the same time, though, he calls the region the hunting grounds of "the Osage, the Creek, the Delaware, and other tribes . . . the Pawnees, the Comanches" where they catch wild horses and hunt elk and bison that still roam "in all their native freedom."[26] The communities he names use the lands regularly and depend upon the Great Plains region for their hunting grounds, but they erect no permanent habitations. In his reading, therefore, they do not fully "inhabit" the country. Irving goes so far as to say that his party is the first Euro-American group to travel in that particular area, while decidedly not the first people.[27] Already, Irving is ambiguous about his party's movement into the Indian Territory. Additionally, he will come to argue that whatever animals are to be harvested for their meat there should be largely domestic, rather than hunted wild, as in the case of the bison. While Irving writes thoughtfully of the tribes he comes across during his brief sojourn in and around the Pawnee hunting grounds, in the end he returns to the predominant idea of his time in the United States, namely, that settlement required permanent habitation.

In "Frontier Scenes," Irving describes a village where U.S. farms exist alongside Creek villages, "the inhabitants of which appeared to have adopted, with considerable facility, the rudiments of civilization, and to have thriven in consequence. Their farms were well stocked and their houses had a look of comfort and abundance."[28] The Creeks, then, are in this scene becoming similar to U.S. Americans, replacing their traditional foodways with a Euro-American agricultural standard. Contrasted to these assimilated Creek farmers, Irving next describes a frontier "settler or squatter," one who is "a tall raw boned old fellow."[29] This man seeks violent retribution for the loss of his horse, assuming it has been stolen by a band of Osage. Indeed, Irving references "Lynch's Law" in his discussion of this episode. Irving's party attempts "to calm the old campaigner of the prairies, by suggesting that his horse might have strayed into the neighboring woods; but he had the frontier propensity to charge every thing to the Indians."[30] Irving makes it clear in these contrasting scenes—placed adjacent to each other in the text—that his readers are to approve of the Creeks rather than the squatter. The Creeks have come closer to using the land in a manner considered appropriate to U.S. agri-expansion in this case, creating farms and building comfortable houses. The squatter, meanwhile, is more violent and paranoid; he has not made such "civilizing" strides as the Creeks. Irving describes the West as an engaging, dynamic world—while at the same time a space available for the national project of agri-expansion under the aegis of Manifest Destiny.[31]

Irving considers his first encounter with the bison—he uses the term "buffalo" throughout—in terms of the hunt: "The Buffalo stood with his shagged front always presented to his foe [Irving's party], his mouth open, his tongue parched, his eyes like coals of fire and his tail erect with rage; every now and then he would make a faint rush upon his foe, who easily evaded his attack, capering and cutting all kinds of antics before him."[32] In this early scene, Irving emphasizes the wildness of the animal, using terminology that expresses danger, size, and intensity. Eventually, Irving joins his party's hunt, just in time to note that they had made a "butchery" of it. They severely wound the animal, rather than effecting a quick kill, causing the bison significant pain before finally "putting the poor animal out of his misery."[33] Irving's hunt is wasteful. Moreover, this "grand" animal dies while standing in water, away from the party, and its meat must be abandoned—further emphasizing its waste.[34] Elsewhere in *A Tour on the Prairies*, Irving writes of the bison as a "poor animal," adding, "His very size and importance, which had before inspired me with eagerness, now increased my compunction. It seemed as if I had inflicted pain in proportion to the bulk of my victim, and as if there were a hundred fold greater waste of life than there would have been in the destruction of an animal of inferior size."[35] Irving describes the bison as a victim, a term usually reserved for humans. However, he is not wholly anthropomorphizing the bison; rather, he specifically describes the creature as an "animal," though one that does not deserve the death it receives via his hunting party. If Irving has not given the bison full personhood, he has allowed it a great symbolic value through his treatment of its death. He understands that this killing is somehow more dreadful than other hunting experiences—perhaps of deer or even of boar—as he would have experienced them in Europe and the U.S. East. Moreover, in his comparison of the size of an animal and its relative worth (i.e., a non-wasteful killing), he gives a regionalist commentary. Irving ironically makes an argument for the introduction of cattle to the region in this scene. Bison are not suited to the agri-expansion to come—they are in fact too wild. Cattle, to Irving as well as his readership, are not in the same category of wildness as bison—they are an acceptable form of beef. Therefore, even as Irving laments the killing of the representative animal of the Great Plains, he also provides a rationale for its replacement and wide-scale destruction.

Irving continues this work in his journals, as well. On September 29, 1832, he notes that he "dined Buffalo meat—rich."[36] A few days later, on October 1, he adds, "Dinner at Mr Austins—Boys at table on one side—girls the other—comp[an]y in centre—rich beef—beautiful honey—cakes—vegetables."[37] In these passages, bison (first example) and cattle (second example, presumably) are both described as "rich." There is a direct connection between the two animals, and here Irving provides a specific example wherein cattle can readily replace bison in the cuisine of the Great Plains. Indeed, both are presented by a term that has economic overtones, "rich." Settlers expanding west have his encouragement to transport their cows with them, it seems, and the Great Plains region has been proven to support their kind. Irving's writings trace what would become an ongoing theme in U.S. culture—the decimation of native animals in place of non-native domestic livestock—as part of the expansion of U.S. agriculture intensifying in the nineteenth century.

During this time, the American bison was sentimentalized in a similar manner to Native Americans—seen as part of a "vanishing wilderness" myth that would be explained by Francis Parkman in the *Knickerbocker Magazine* in 1849. The last bison of the East, a cow and her calf, had been killed in West Virginia in 1825.[38] Henry Nash Smith observes that Parkman lamented the "doomed" wilderness: "Cattle would soon replace the buffalo and farms transform the range of the wolf, bear, and Indian."[39] Bison, having become part of the image of the Great Plains, were therefore already threatened by this time. Irving writes to his brother of his desire to see these bison, as well as the Native American tribes of the region, "while still in a state of pristine wildness."[40] While environmental elements added to the decimation of the bison, more important was the fact that U.S. hunters were encouraged to slaughter millions of bison by federal authorities who equated their removal with political power over Great Plains nations and tribes—part of ongoing efforts to force them onto reservations. Indeed, from first contact through the twentieth century, bison numbers dropped from as many as twenty-seven million to only a few hundred individuals, as Charles C. Mann notes. Root and de Rochemont add that this destruction began in the 1830s, the time of Irving's travel writings. In 1840 alone, for example, St. Louis collected the skins of enough bison for "67,000 buffalo robes."[41] The bison were almost as quickly being replaced with domestic cattle—animals that traveled with U.S. settlers to the territories west of the Mississippi, part of the nation's larger agri-expansionist plan to establish massive domestic herds across the West. The bison became utterly doomed as part of the development of the United States and its form of animal husbandry into the Indian Territory and other U.S. territories, and Irving encounters an early element of this in *A Tour on the Prairies*.

Toward the end of *A Tour on the Prairies*, Irving's party again approaches a frontier habitation, a Creek village complete with log houses and "small farms," but they continue farther, seeking "the habitation of a white settler" to beg food from instead.[42] The party does not consider the Creeks, apparently, civilized enough culinarily for their purposes. Shortly thereafter, Irving finds a more suitable farm: "Here was a stable and barn and granaries teeming with abundance, while legions of grunting swine, gobbling turkeys, cackling hens and strutting roosters swarmed about the farm yard."[43] He does not describe cattle in this scene, though; the cows might be out to pasture, or the settlers might not have the funds to afford them yet. In this way, these frontier farmers resemble those that I will describe in more detail in chapter 6, with Charles Chesnutt's discussion of agriculture in the U.S. South as a relatively cattle-free region. However, beef does make an appearance in this scene. The first foodstuffs Irving sees in his hunger is a stew: "a broad and smoking dish of boiled beef and turnips."[44] The farmer's wife, who, ironically, is not Euro-American but rather African American, provides him the much desired food: "Placing a brown earthen dish on the floor she inclined the corpulent cauldron on one side, and out leaped sundry great morsels of beef, with a regiment of turnips tumbling after them, and a rich cascade of broth overflowing the whole . . . and then such magnificent slices of bread and butter! Head of Apicius, what a banquet!"[45] This rough farmhouse has become Irving's ideal, existing on the frontier. The missing cattle are presented ready-dressed and prepared to eat. This scene shows Irving's idea of the best use of the Great Plains—the most civilizing distance from the land of roaming bison that might be improperly slaughtered and even wasted. This exemplary farm contains cows that are ready to be eaten almost instantly. Irving's progress throughout the narrative of *A Tour on the Prairies* foreshadows the nation's movement through the prairies over the rest of the nineteenth century, as he envisions a future of U.S. agriexpansion, even while providing a nostalgic tourist's adventure narrative. Following his depictions, and others like his, the bison would be replaced over the course of the nineteenth century by the domesticated cow.

At a key point in his narrative, Irving makes a direct suggestion to future U.S. settlers to convert the Great Plains from bison to domestic cattle: "We descried a herd of buffalo about two miles distant, scattered apart and quietly grazing near a small strip of trees and bushes. It required but little stretch of fancy to picture them so many cattle grazing on the edge of a common and that the grove might shelter some lowly farmhouse."[46] Irving witnesses an image much like one represented in a painting by George Gatlin (see fig. 3). In this way, Irving makes his domestication project of the prairies, through a replacement of bison with cattle, explicit. What is implicit in this description is the removal of Native Americans by Euro-American settlers involved in this plan. Following Irving's vision, they would cohabitate—even in a limited manner—no longer. This apocalyptic vision of Irving's is alarmingly considered only a "little stretch of fancy." According

Fig. 3. *Elk and Buffalo Grazing among Prairie Flowers, Texas,* 1846–48. The iconic painter of the West, George Gatlin, here portrays an idyllic scene of wild elk and bison in the Great Plains region. Smithsonian American Art Museum, gift of Mrs. Joseph Harrison Jr.

to John Schlueter, Irving's idea of "fancy" in his writings in *The Sketch Book* "indicates a process of thinking governed by association . . . a mental wandering, or a kind of day-dreaming."[47] This analysis applies to the scene with Katrina from earlier in this chapter. I argue, as well, that this mental daydreaming remains in *A Tour on the Prairies.* Irving has reduced the work, though, as it requires only "little" of that daydreaming fancy to shift one's vision from bison to cows, Native American tribes and nations to Euro-American farmers. Specifically discussing bison and Irving's "stretch of fancy" scene, Burns comments that Irving's fanciful picture has the bison inhabit "a pastoral, bucolic setting— precisely *not* the kind of natural context he had seemingly come all the way to the western frontier to see."[48] While Burns says that the failure to achieve authenticity on the frontier is inherent to the concept of the U.S. frontier itself, I maintain that Irving succeeds wildly in the sense that U.S. settlers and farmers will shortly flock to the frontier in greater numbers than Irving or anyone else at the time could have imagined.

Irving visited the Great Plains, also called the Great American Desert, at a transitional time, from the reign of the bison and the Native American tribes who depended on them, to that of domestic cattle and Euro-Americans; indeed, Irving helped narrate this transition. In his account of the bison hunt, Irving laments the changing prairie lands even as he documents and recommends those very changes that would make it the breadbasket of the world and the land of feedlot cattle over the second half of the nineteenth century. Irving's western writing was an appeal for agri-expansion across the continent. Like the agricultural expansionists who pushed into the Indian Territory throughout the nineteenth century, Irving combined economics and literature in his own form of expansion. In his western writings, Irving imagines an economic boom, perhaps something connected to the cattle kingdom of the late nineteenth century, where others lamented a fear of the lack of arable lands in the Great American Desert. Indeed, Irving's hopeful imaginings bore national fruit. His writings would help spur a new wave of territorial expansion that would last through the second half of the nineteenth century. His example, combined with the writings of other popular authors of the early nineteenth century, proved irresistible to U.S. audiences. At the same time, Irving's narrative complicated the assumptions within U.S. agri-expansion: the Creeks were assimilating successfully to U.S. agriculture, U.S. squatters were sometimes shameful examples of frontier farmers, and his ideal farm was run by a biracial couple. Finally, though, Irving's underlying message prevailed: that the Great Plains were being used wastefully by the bison, which were then wastefully hunted. Irving convincingly wrote in favor of expansion of livestock husbandry farther west. He was a romantic in terms of his portrayal of the prairies, and through his descriptions of the country—and its plentiful land—he encouraged settlement in the region. This expansion, the questions of contested ownership and consumption of lands and resources, and the role of domestic cattle within agri-expansion would continue to be themes of Irving's other western narratives, *Astoria* and *The Adventures of Captain Bonneville*.[49]

The Cattle-Free West of *Astoria*

As Irving describes his own experiences west of the Mississippi in *A Tour on the Prairies*, his next work, *Astoria*, narrates the adventures of others. The 1810 Astoria Expedition to Oregon was funded by the first U.S. millionaire, John Jacob Astor, who made his fortune on furs as founder of the American Fur Company. Astor desired a Pacific fur empire, to further trade with China and other Pacific Rim nations. In 1810 he sent two parties—land and sea—to the Pacific Northwest, an area contested at the time between Britain, Spain, Russia, and the United States. Astor's expedition parties eventually arrived at Fort Astoria in 1812, only to become embroiled in the politics of the War of 1812, and the region was for the time lost to the British. The Astorians returned east by land, arriving in St. Louis, Missouri, in 1813. Still viewing the experience positively, Astor commissioned Irving, one of the nation's most popular authors of the time, to publish an account of the experiment. Irving articulated the West as a land of economics and innovation. His writing foreshadows Michael Hardt and Antonio Negri's description of contemporary imperialism as "decentered" if not "deterritorialized."[50] In his narratives, Irving navigated between idealized descriptions of Edenic valleys and the decentered markets in play in the Pacific Northwest in the early nineteenth century. In so doing, he helped describe what would become the U.S. Northwest. *Astoria* was based on events that predate his western tour described in *A Tour on the Prairies*. In this western narrative, readers are introduced to the West before the introduction of cattle and other domestic animals by Euro-American settlers. I argue that Irving shows the benefit of introducing cattle to the Northwest in key scenes of *Astoria*, benefits that would have lasting impact on U.S. agri-expansion.

In Irving's narrative, the West is anything but domesticated in 1810. One of the most remarked upon passages of *Astoria* occurs when Irving theorizes the future of the West: "Some portions of it along the rivers may partially be subdued by agriculture, others may form vast pastoral tracts, like those of the east; but it is to be feared that a great part of it will form a lawless interval between the abodes of civilized man, like the wastes of the ocean or the deserts of Arabia. . . . Here may spring up new and mongrel races."[51] Irving does not look as hopefully on the growth of the agrarian model in the Northwest as he had for the Great Plains of *A Tour on the Prairies*. There is a chance, he observes, for successful agriculture in this instance—something comparable to the familiar scenes of the established East. However, he "fears" it might be more likely to become another desert like Arabia—or an ocean waste—and "lawless." He anticipates something that approximates the "wild west" more than the frontier farms he had envisioned in *A Tour on the Prairies*. Indeed, it might be the lack of domestic animals that leads to his "mongrel" reading of the West that threatens to un-civilize the course of U.S. expansion.[52] However, the Pacific Northwest was already "mongrel," or multifaceted, at the period in which Irving was writing. In the fur-trading industry along the Pacific Northwest, the nations represented are European, American, and Asian, leading up to and following Indian removals and John L. O'Sullivan's naming of Manifest Destiny. However, during the Astoria enterprise "this transformation of land into property, and the Indian into illegitimate holder of that property," Peter Antelyes writes, "was a fundamental principle in the rhetoric of expansion" in the contested sites of the Pacific Northwest.[53]

Astoria combines literature, through Irving, the "era's most celebrated writer," with capitalism, through Astor, the "most celebrated capitalist," as described by Peter Jaros.[54] Irving's narrative ends with the failure of the original Astoria expedition, though the Oregon Treaty of 1846 would ultimately allow the United States and Britain to settle borders between the U.S. Northwest and Canada. I agree with Jaros's observation that *Astoria* "is usually read, when it is read at all, as a lament for an unrealized empire and a call for further US expansion."[55] By the 1830s and 1840s, indeed, immigration would explode into the Pacific Northwest, and Irving's writing was an influence on that explosion. Irving's narrative garnered interest in a northwestward expansion than ran somewhat parallel to the gold and silver rushes of the late 1840s through the 1860s in California and Nevada and, later, Alaska. The territory involved an economic network across New York (Astor) and the Pacific Northwest (Fort Astoria) as well as London, eastern Canada, the South Pacific, Russia, and China. The Pacific Coast was part of both regional and global economies in the early nineteenth century. If originally the United States looked for an established location to engage with this transnational fur trade, the purpose for settlement would change as frontier farmers started to establish themselves in the West.[56]

As shown in the journals of the Lewis and Clark Expedition, one of the most noteworthy elements of Irving's use of domestic cattle in *Astoria* is that there are, indeed, no cows present yet in the Pacific Northwest. Euro-American colonization had not occurred in the West by 1810, excepting in the lands of Mexico—what would become the U.S. territory from Texas to California—where domestic cattle were already being raised.

Instead, there were bison in the Great Plains, and not even that form of bovine west of the Rockies. Irving describes the lack of cattle, and the idea of bison-as-cattle, as he imagines it to have been for the land-based Astoria expedition as they crossed the prairies.

Boundless wastes kept extending to the eye, more and more animated by herds of buffalo. Sometimes these unwieldy animals were seen moving in long procession across the silent landscape; at other times they were scattered about, singly or in groups, on the broad enameled prairies and green acclivities, some cropping the rich pasturage, others reclining amidst the flowery herbage; the whole scene realizing in a manner the old scriptural descriptions of the vast pastoral countries of the Orient, with "cattle upon a thousand hills."[57]

Bison were plentiful in the Great Plains in 1810; however, eastern readers would not have been as familiar with them as they would have been with domestic cattle, even in the 1830s in which Irving wrote. Indeed, a *Massachusetts Magazine* article of 1792 would include an engraving and description of the "buffalo" for a curious readership, giving the animal seemingly human features (see fig. 4).[58] Irving therefore turns to common cultural symbols to describe the bison to his readers: cattle in the Bible. By connecting these two motifs, Irving is able to express some of the size and scale of the bison in the region. Irving's version of "buffalo" is, in fact, described like domestic cattle. They are not stampeding, but rather grazing in silent processions. They are portrayed in an idyllic manner—including the biblical reference to the "pastoral." Cattle might not be physically apparent in the Northwest of 1810, but the bison are cow-like in their numbers and, apparently, in their ability to become domesticated. Irving takes his imagery even farther east than the United States in his description of these biblical cattle—to the east of Arabia, his

Fig. 4. "The Buffallo," *Massachusetts Magazine*, 1792. In the early nineteenth century, residents of the eastern United States who were unfamiliar with bison would have turned to artistic representations like this image. Note the anthropomorphized features of the bison. Courtesy of the American Antiquarian Society.

earlier "mongrel" land, and the "vast pastoral countries" to be found there. In contrast, Irving uses cattle to show a progression followed by Euro-Americans upon colonization, one that Irving would call upon for more lawful expansion of U.S. agriculture into the Pacific Northwest. David Watson compares Irving's narrative to Jackson's political project in the 1830s: "To speak within this frame of the frontier as an English park or a European scene is to rhetorically domesticate and cultivate it, and prepare it for settlement."[59] That is precisely what Irving does in his later travel narrative, *The Adventures of Captain Bonneville*. He imagines a settled U.S. West—including domestic cattle—as opposed to its existing form, which centers around bison and the Native American nations and tribes dependent upon them.

Cattle Expansion in *The Adventures of Captain Bonneville*

In *The Adventures of Captain Bonneville*, therefore, Irving engages with the developing U.S. West from a third perspective, that of an individual adventurer, rather than an expedition of hopeful settlers or a party of hunters. While researching his narrative for Astor, Irving was introduced to a U.S. Army officer, Captain Benjamin Bonneville, who had received military leave from duty in 1832 to explore the Rocky Mountains and regions farther west on the continent. His expedition traversed the Oregon region as well as the Great Basin and California, ultimately returning east in 1835.[60] Irving's narrative of Bonneville and his party has more cattle-related scenes than does *Astoria*. These scenes, interestingly, largely involve cross-cultural interactions between the expedition, Native Americans, and Californio/a vaqueros. Early in Bonneville's narrative, for example, members of the Crow nation begin to act strangely upon seeing the U.S. party, who then become alarmed that the Crows might mean to do violence to them. However, that proves to be a misreading of the situation; indeed, the Crows mean violence only to "a band of Cheyennes" against whom they are fighting. The U.S. explorers at that point are not involved with the politics of the battling Native American nations. In the midst of this moment of miscommunication, though, it becomes apparent that the Crows have never before seen domestic cattle, and have started to follow the wagon train to get the proverbial closer look. They are "especially struck with the sight of a cow and calf, quietly following the caravan; supposing them to be some kind of tame buffalo."[61] This is such a dramatic view that the Crow scouts return with their chief, who "diverged from his pursuit of vengeance to behold the wonders described to him."[62] They are profoundly curious about the new animals, as they make clear to Bonneville's entire party: "Waggons had never been seen by them before, and they examined them with the greatest minuteness; but the calf was the peculiar object of their admiration. They watched it with intense interest as it licked the hands accustomed to feed it, and were struck with the mild expression of its countenance, and its perfect docility."[63] This singular transcultural view of the domestic cattle, the "tame buffalo" and its calf, contextualizes to the Crows some of the wonders of U.S. agriculture—the ability to make the wild bison tame. They cannot understand all of what would come through the next several decades in the west of the continent, or presumably they would not have proclaimed upon this scene that "our hearts are glad."[64]

These same Crows assume incorrectly that there must be a "supernatural and mysterious power" or "great medicine" that the Bonneville party possesses in order to give them power over these tame bison.[65] However, the party is stymied in their efforts to manipulate this situation when the Crows reject their offer "to exchange the calf for a horse."[66] In this instance, it seems as though both the U.S. adventurers and the Crows value horses more than domestic cattle—especially dependent cattle before they are weaned. Horses had well-established value for Euro-Americans and Native American nations like the Crow nation, while cows were not known yet to be as obviously important. Neither party realizes that the calf is truly the champion of the U.S. West, as it represents domestic animals, agriculture, and U.S. settlement. To some extent, this is a primal scene for the meeting of cultures, foreshadowing what would come over the rest of the nineteenth century.

The Bonneville party shortly thereafter realizes that cattle have already been introduced to certain parts of the West, farther south than Oregon. In a later section of the narrative a subgroup of the Bonneville party ventures to California, for reasons historians have been speculating about ever since, as it went beyond the scope of the party's project and the Oregon Trail. This expedition brings domestic cattle into Irving's narrative in another form; this time it is the Spanish-Mexican longhorn cattle being represented. Irving introduces these longhorns and the Californios/

as and Native American communities of California, the "vaque-ros, or Indian cattle drivers" who work with cattle.[67] First, Irving glosses over the historical abuses of the mission system of slavery, simply noting that perhaps the Jesuits "have generally proved the most beneficent of colonists," as opposed to the Franciscans under Father Serra.[68] In either case, the Native Americans of the region "are taught husbandry, and the rearing of cattle and horses."[69] In truth, they are forced to learn a different style of agriculture, including animal husbandry. This Spanish-Mexican method involves the rearing of cattle: "Horses and horned cattle abound throughout all this region," Irving writes.[70] Meanwhile, though the Native Americans in Irving's narrative have not had the choice of whether or not to become vaqueros, they have come to excel at it. For example, Irving describes their skill with the lasso to his readers. The vaqueros "employ it [the lasso] to catch the half-wild cattle, by throwing it round their horns."[71] This scene would go on to become one of the most representative images of the U.S. West: the cowboy lassoing a "half-wild" steer. In his description of it, interestingly, Irving indicates that the cowboys here are the Native American tribes of California (a theme that will be discussed further in chapter 5). Irving also references the Bonneville party marveling at the "equestrian skill of the Californians."[72] Both the Bonneville party and the Californios/as esteem horses, understanding their usefulness in the West. Therefore, for the easterners to marvel at the skills of those they see in California, Spanish-Mexican as well as Native American vaqueros, is especially significant. Irving presents his readers with an early vision of the U.S. cowboy in his description of the cattle drivers of Monterey, California, and in so doing he once again popularizes the West for his readers.[73]

In addition to influencing U.S. settlers, Irving's writing had power over contemporary politics, as well. In one example, U.S. congressman Caleb Cushing used Irving's narratives to further his personal political goals. Cushing's foreign policy involved the Treaty of Wanghia (1844), advocacy of the Mexican-American War (1848), and the Oregon Treaty (1846) with England, which gave the territory, including Astoria, to the expanding United States. Indeed, Cushing publicly discussed *Astoria*—as well as Irving's other western writings—in terms of its "romantic verve" as an actual tool used in Congress to fight for U.S. expansion into the West; thus, Irving's concern about the nation's ability to apply a form of agri-expansion as an extension of Jeffersonian agrarianism was being seen in Congress itself. Whether it was Irving's intention or not, his words directly influenced policy makers in the first half of the nineteenth century in terms of U.S. agri-expansion into the West for the purposes of cattle production. As seen with these U.S. policy makers, Irving prolongs his literary influence throughout his career. If the 1820s brought him international celebrity through *The Sketch Book*, then the 1830s brought him national celebrity through his narratives of U.S. expansion into the West. Throughout, he used cattle as a cultural focal point between Native American peoples and U.S. settlers. Domestication and the West were literalized in Irving's own life. His western narratives were domestic and expansionist at the same time. Cattle made up an important part of his overall contribution to U.S. expansion as it had with U.S. nationalism earlier in the century. Indeed, Irving largely popularizes the nation's attitudes to westward expansion. The disastrous effects that movement had on the Native American nations and tribes throughout the West will be discussed in more detail in chapter 4. This period of Irving's writing career was associated with the promotion of that expansion—as part of a move to bring cattle west.[74]

Conclusion: Sketching the West

Returning to his most famous stories in *The Sketch Book*, I argue that Irving begins to contemplate ideas about movement, contact, wealth, and the potential of expanding west already in these stories, using symbolic domestic animals to do so. That Irving chooses cattle as a symbol through his writings on the United States and specifically the U.S. West was not out of character for him. He used an impressive number of food-related scenes and references throughout his stories and narratives. In "The Legend of Sleepy Hollow," Ichabod Crane compares everything he encounters directly to foodstuffs. Whether he is dressing the domestic animals on the Van Tassel family farm into delicious foods in his mind or imagining Katrina Van Tassel as a plump partridge, the theme of food—specifically livestock—carries throughout the story. Food scenes appear throughout *The Sketch Book*, and they continue in his western narratives as well—as part of the consuming drive for territory in which to raise livestock. Key scenes from these sketches would be mirrored in his western narratives involving Daniel Boone–style characters like the ironic Ichabod Crane or the impressive Captain Benjamin Bonneville. Richard Slotkin maintains that the real Daniel Boone also compared cattle to bison in his writing, much as Irving does: Boone writes, "The buffaloes were more frequent than I have seen cattle in the settlements."[75] In both cases, cows are used in part as a symbol to explain bison to an unfamiliar readership.[76]

Irving's popular narratives caused a dramatic shift in the national culture, specifically, as settlers began moving west in throngs even before the California gold rush. Biographer Brian Jay Jones calls Irving "an American original."[77] A timely cross between U.S. writers and personalities like Benjamin Franklin and Mark Twain, Irving became an icon in U.S. literature through his humorous, satirical, and historical works. He "perfected" the U.S. short story—influencing generations of authors, including Charles Dickens. Irving's *The Sketch Book* influenced *Sketches by Boz* (1836) and *The Pickwick Papers* (1837) specifically. Indeed, Dickens would make a tour of the United States commemorated in his 1842 *Amer-*

ican Notes. In it, though, Dickens expresses a decidedly negative judgment of U.S. cuisine where native foods are privileged and beef is cheap, plentiful, and seemingly prepared without style. He describes a train journey through its meals: "Although there is every appearance of a mighty 'spread,' there is seldom really more than a joint: except for those who fancy slices of beet-root, shreds of dried beef, complicated entanglements of yellow pickle; maize, Indian corn, apple-sauce, and pumpkin."[78] He dislikes the food items that were most identified with the United States, like Indian corn, the pumpkin, and even the plentiful—if not native—beef. Regardless of Dickens's disapproval in this instance, Irving's influence on a U.S. audience in the nineteenth century can hardly be overstated, and it continues to this day. Regional commonplaces remain that are credited to him, like the term "Gotham" for New York City, the New York Knickerbockers basketball team, and the common "Sleepy Hollow" neighborhoods located across the U.S. Northeast. The idea of studying the foods of the expanding nation would continue into the middle of the nineteenth century, after the time of Irving's last writing, *The Life of George Washington*, "his final act of conscious mythmaking written to hold back the night of that most dreadful of America's political decades, the 1850s," as Jerome McGann writes.[79] Irving wrote of U.S. culture and territory for much of the first half of the nineteenth century, while living at home and abroad, seen especially along the contested spaces of the expanding western frontier. As a comparison to Irving's writing project, in the following chapter I analyze Irving's contemporary James Fenimore Cooper and Cooper's daughter Susan Fenimore Cooper as they challenge expansion more directly than Irving in their writings of domestic animals and the frontier. Like Irving, they privilege a specific kind of agriculture, but they spend a greater amount of time than Irving does in warning their audience about the hazards to the environment that might result from examples of intensive agricultural practices.

TWO

Civilizing Cattle in the Writings of James and Susan Fenimore Cooper

And when the proud forest is falling,
To my oxen cheerfully calling,
From morn until night I am bawling,
Woe, back there, and hoy and gee;
Till our labour is mutually ended,
By my strength and cattle befriended,
And against the musquitoes defended,
By the bark of the walnut tree.—
Away! then, you lads who would buy land,
Choose the oak that grows on the high land,
Or the silvery pine on the dry land,
It matters but little to me.

—JAMES FENIMORE COOPER, *The Pioneers*

In this song from James Fenimore Cooper's *The Pioneers* (1823), the secondary character Billy Kirby describes his oxen, his "cattle befriended," as they help him labor as a woodcutter in the steadily depleting forests around the town of Templeton, a fictionalized version of Cooperstown, New York, along the early U.S. frontier. Kirby's oxen help him cut down trees and clear the region's forests more quickly, thereby making more pastureland available for increasing numbers of domestic animals—like beef and dairy cows. Kirby connects one form of cattle, in this case oxen, with deforestation and a specific kind of Euro-American animal husbandry. In his song, Kirby doesn't care if he clears walnut, oak, or pine; he sings that all are equally disposable for him in the interest of bringing what I have termed agri-expansion to the region. Indeed, Kirby opposes Judge Marmaduke Temple, one of Cooper's protagonists in the novel, in his ideas of forest conservation. Kirby says of the forest's trees, "To my eyes, they are a sore sight at any time, unless I'm privileged to work my will on them. . . . Stumps are a different thing, for they don't shade the land; and besides, if you dig them, they make a fence that will turn any thing bigger than a hog, being grand for breachy cattle" (see fig. 5).[1] Kirby describes precisely what the chopped trees are to be used for: fences for domestic cattle. While Judge Temple laments the signs of deforestation, a woodchopper like Kirby enthusiastically toils at extending it. The conflict between these two philosophies of resource management in the interest of U.S. agriculture appears as a prominent theme in the writings of James Fenimore Cooper as well as those of his daughter Susan Fenimore Cooper.[2]

In *The Pioneers*, James Cooper shows scenes where the land's gifts are inappropriately used, like the wasteful hunt of passenger pigeons lamented by Leather-stocking, in addition to the deforestation described above. These episodes typify the concerns shown by both Coopers, father and daughter, in their other works, *The Prairie* (1827) and *Rural Hours* (1850), respectively. Throughout these writings, the Coopers examine methods of appropriate resource management, lament environmental degradation, and speculate on the future of wild animals—like the passenger pigeon and American bison—as opposed to domestic livestock. In each case, "civilization," specifically Euro-American agriculture, brings about anxieties in terms of the use of land, in

Fig. 5. In James Fenimore Cooper's writing, Billy Kirby would have created stump fences like this one, from 1890, in his work as a wood-chopper. These were used during the nineteenth century throughout the New England region as well as Upstate New York. Library of Congress, Prints & Photographs Division, Detroit Publishing Company Collection.

addition to the political issues of navigating land rights with the indigenous communities already inhabiting them. Both Coopers believed in the superiority of Euro-American agricultural practices over those of Native American nations and tribes. However, they also argued that there were other valid forms of knowledge available across the vast continent inhabited by multiple indigenous communities. Both Coopers engage directly with issues of race and the environment in their writings on the developing nineteenth-century frontier, therefore. With James Cooper's startling ending of *The Prairie,* where U.S. squatters return to the East following their disastrous efforts to become farmers in the West, and Susan Cooper's romanticization of local agriculture and environment in Cooperstown, New York, both father and daughter present arguments against excessive expansion. Rather, they emphasize a reverence for the local, specifically noteworthy in Susan Cooper's nature writing. Rochelle Johnson writes of three definitions of "nature" significant during this period in U.S. literature: "1) nature as national progress; 2) nature as the refinement of the American people; and 3) nature as human reason."[3] I focus primarily on the first two "meanings" of nature in the writings of the Coopers. Nature is both refining and a means of progress that the Coopers engage with, agree with, and occasionally question throughout their writings. In this chapter I examine scenes in *The Pioneers* and *The Prairie* as well as in *Rural Hours* where the Coopers show their concerns about the developing culture of the U.S. frontier, dealing with Native American and Euro-American relationships as well as the appropriate use of environmental resources.

James Cooper was a contemporary of Washington Irving, and Susan Cooper's nature writings predated and likely influenced Henry David Thoreau, whose work I will examine in the following chapter.[4] As Irving did in his stories and sketches about Upstate New York, the Coopers combined Leo Marx's "middle landscape" with the frontier concept of Frederick Jackson Turner into their own form of Jeffersonian agrarianism. Turner specifically describes the Cooper family in his history, *The United States.* The Coopers were part of a lineage of settlement that dated back to Euro-American colonization. Since that time, Euro-Americans had used domestic animals to help European settlement on the continent. For example, oxen, mules, and horses would help plow the newly cleared fields, making the land ready for the grazers—sheep, goats, and especially cattle. In this way, livestock, oxen and other cattle in particular, became a catalyst of expansion affecting early Native American and Euro-American relations, and this history would be repeated into the nineteenth century along the expanding western U.S. frontier. During the century, the United States expanded to new lands in order to have space to allow domestic cattle to graze—and eventually to grow feed for cattle in large-scale operations. The livestock aided in expansion, as well as benefiting from it. James Cooper wrote his first Leather-stocking tales before the expansionist era of Andrew Jackson, but the figure of Leather-stocking foreshadows arguments against such expansionism and what it would do to Native American and Euro-American communities.[5] While both Coopers would consider U.S. agriculture as "civilized" and Native forms of agriculture as "savage," they regretted the cost to Native American cultures as well as the environmental damage done through deforestation and the loss of species diversity that they witnessed throughout their lives as a consequence of such agri-expansion. Their argument for greater caution and stewardship in the development of villages along the frontier for the use of Euro-American farmers appears throughout their novels and nature writings, I argue, as a precursor to modern ecocriticism.[6]

The farming history of the Coopers, then, helps to explain their interest in small-scale agriculture at a time of U.S. expansion. William Cooper, James Cooper's father, developed trade routes across the port cities of the East Coast in order to "develop Cooperstown's land-based production, not only cattle and crops, but also minerals (failures), maple sugar (largely a failure), and potash (a success)," according to Timothy Sweet.[7] Livestock are mentioned first in this description; they are neither a success nor a failure, but rather an assumed presence foundational to any further economic endeavors. Later, James Cooper would raise his own family in part on his beloved Fenimore Farm. Ironically, the farm primarily raised merino sheep rather than cows, the

Coopers having procured 145 sheep within their first year of operation, taking advantage of international markets in wool.[8] Cattle were of course an important part of the Coopers' farming experiment as well, and both father and daughter would write about the cattle of Fenimore Farm in family letters. Susan, James's second daughter and the first to survive infancy, would develop her theories about agricultural stewardship in her nature writings from the perspective of the family farm. In addition, she was directly connected to the family business of writing and publishing as her father's amanuensis; the two Coopers worked together for years, she editing his work and promoting his writings after his death, and he helping her to publish *Rural Hours*. Thus, their interests in both literature and agriculture were closely aligned. The Fenimore Farm was a significant theme for the Coopers as they tried to maintain it, worried about losing it, tried to repurchase it, and eventually (after James's death) gave it up forever. It even helped to establish the particular significance of Cooperstown to U.S. culture. As their descriptions of regional Cooperstown influenced artistic culture during their lives, it has also had a profound significance on the Coopers' readers ever since.[9]

Environmental and agricultural themes are at the heart of James Cooper's writing, seen in the example above from *The Pioneers*. The plot of the novel revolves around Leather-stocking's illegal killing of a deer. Leather-stocking is eventually arrested and jailed for the crime of "killing of deer in the teeming months."[10] Although he is deemed a criminal by Judge Temple's laws encouraging conservation of nature, Leather-stocking also poses legitimate questions about who truly owns the deer and other fruits of the land. Cooper connects the ability to hunt, unimpeded by the laws established so proudly by Judge Temple, with the ability for Chingachgook, or John Mohegan, a friend of Leather-stocking's and the eponymous character from *The Last of the Mohicans* (1826), to make a living. Thus, critical race studies also lie at the heart of Cooper's novels. Chingachgook represents Cooper's romanticized vision of northeastern Native American cultures, an ideal that would have been in line with most of Cooper's nineteenth-century readership. Chingachgook and Leather-stocking repeatedly draw attention to the hypocrisy of Euro-American civilization. After the two kill a deer in flagrant violation of Judge Temple's law, Cooper writes, "The Indian had long been drooping with his years, and perhaps under the calamities of his race, but this invigorating and exciting sport caused a gleam of sunshine to cross his swarthy face, that had long been absent from his features."[11] Northeastern nations and tribes assimilated to the agricultural methods of Euro-American settler colonizers as much as possible. Their agricultural reformations are just some of the "calamities" that Cooper references here. He claims that Chingachgook to some extent needs to hunt in order to maintain his health, even in his old age; he describes this way of life, that of a hunter, as an essential part of him. In this way, Cooper oversimplifies the experiences of the numerous northeastern Native American nations and tribes. However, he also draws his readers' attention to some of the threats of U.S. expansion that were linked intimately with a particular vision of agriculture and the environment.

The intersection, then, of theories of race and environmentalism provides a valuable lens to interrogate the Coopers' writings. Michael Omi and Howard Winant's theory of racial formation describes the "centuries-long conflict between white domination and resistance by people of color."[12] In this manner, the conservatism in the Coopers' writings connects directly to national and cultural policies toward Native American communities throughout the nineteenth century. Meanwhile, nature, wilderness spaces, and middle landscapes dominate the settings of the Coopers' work. In terms of linking the matters of racism and environmental degradation, Jeffrey Myers observes, "Just as the formation of whiteness in opposition to the racial Other is a construction with no basis in the natural world, the formation of the Western, individual, subjective self in opposition to nature is an equally fictional construction."[13] I agree with Myers's argument here, and I focus on how these constructions were particularized in U.S. agriculture during the first half of the nineteenth century. Furthermore, while James Cooper remains a sometimes problematic representative of his period, as seen in his dealings with the themes of race and the environment, in his writings he decidedly favors stewardship—the responsible management and preservation of the land—over agri-expansion.[14]

While Leather-stocking fights for his ability to hunt as part of what he considers to be a natural manner of being, Judge Temple boasts of attempting to pass a law "to make the unlawful felling of timber a criminal offence," in addition to the laws he created against hunting out of season.[15] Temple demonstrates James Cooper's philosophy that preserving deer populations and forests matters, suggesting a way to conserve the environment. Even though Judge Temple's laws mean that Leather-stocking and Chingachgook are not able to hunt for their own sustenance, Temple stands by them. The character finds himself drawn to action during an early instance of environmental degradation, and he creates policies in favor of conservation from that point. However, to these laws, Leather-stocking responds, "No, no, Judge, it's the farmers that makes the game scarce, and not the hunters."[16] Leather-stocking, rather than Temple, here seems to voice Cooper's own views on the issue of stewardship, as he here makes the direct connection between Euro-American farmers and the decimation of the wildlife populations in the Northeast. Cooper presents multiple sides of the fraught history of U.S. agriculture and settlement of his time as it relates to resource management in this scene. The Euro-American form of agriculture was connected to one kind of intensive domestic husbandry, the production of animals like sheep, chickens, pigs, and cattle for meat and other products, as well as horses, mules, and oxen for labor.[17] Cooper complicates this by showing both sides to the issue of consumption and conservation, with relatable characters presenting arguments about whether to privilege domestic livestock or wild game in the developing villages of the U.S. frontier. While Judge Temple is supported by the law, Leather-stocking and Chingachgook are sentimental favorites. They are imprisoned for their crime, but they are also eventually allowed to escape. Leather-stocking then makes an interesting move, deciding to leave this "civilization" and agricultural policies of the region around Templeton, moving west, where he will be discovered again in *The Prairie*.

Livestock, as opposed to wild game, also play an important role in Susan Cooper's nature writing of the Cooperstown region. As opposed to wild animals and the frontier of the time of *The Pioneers*, by *Rural Hours* the land has been thoroughly domesticated by the iconic animal of U.S. domestic husbandry, the cow. Early in *Rural Hours*, Susan Cooper writes of neighborly cows returning home from the day's grazing in open pastures:

> Came home from our walk with the village cows, this evening. Some fifteen or twenty of them were straggling along the road, going home of their own accord to be milked. . . . Toward evening, they turn their heads homeward, without being sent for, occasionally walking at a steady pace without stopping; at other times, loitering and nibbling by the way. Among those we followed, this evening, were several old acquaintances, and probably they all belonged to different houses; only two of them had bells. As they came into the village, they all walked off to their owners' doors, some turning in one direction, some in another.[18]

In this scene, the cows accompany Cooper on her walk companionably, almost in a kind of friendship. Their similarity to their human companions is highlighted at the end, as the figures detach from the group. The cows separate themselves one by one, at the appointed time, like a human walking party. This scene establishes to a large part the general tone of the narrative: Cooper and her family are in comfortable conversation with the nonhuman world around them.

Throughout *Rural Hours*, Susan Cooper compares the domesticity of herself and her neighbors on their farms with the threats to this level of comfort: westward movement and agri-expansion. She applies a historical reading to her mid-nineteenth-century depiction of the Cooperstown region—one that speaks to both *The Pioneers* and *The Prairie*. She laments the agri-expansionism that James Cooper describes in *The Prairie*, his vision of U.S. expansionist futures. Indeed, James Cooper's most famous creation, the character Leather-stocking, became a frontier trapper as opposed to a "civilized" member of society in the domesticated Cooperstown described in Susan Cooper's nature writing. He becomes a version of the "pioneer" mentioned in the title of James Cooper's first Leather-stocking novel as he follows the frontier west. Interestingly, this question of agri-expansion, of farming, hunting, and conservation, would become one of the main themes of James Cooper's third Leather-stocking novel, his original intended conclusion to his Leather-stocking series, *The Prairie*.[19]

Civilizing Livestock in *The Prairie*

James Fenimore Cooper's *The Prairie* (1827) tells the story of a family of squatters—those who inhabit the land without legal claim to it—as they move across the desert-like Great Plains and come into contact with various Native American nations and tribes—the Teton Sioux and Pawnee in particular. Moreover, Cooper's Daniel Boone–like Leather-stocking character, called "the trapper" in this novel, reappears as a heroic figure in this sentimentalized account of the western frontier of the expanding nineteenth-century United States. The plot of the story involves the kidnapping and theft of both humans and domestic animals. First, the Bush family—specifically the married couple Ishmael and Esther Bush, and her brother, Abiram White—kidnap the newly married Inez de Certavallos-Middleton for ransom. Inez's husband, Duncan Uncas Middleton, quickly pursues them, making the connection to earlier Leather-stocking novels through his grandparents, the characters Major Duncan Heyward and Alice Munro Heyward from *The Last of the Mohicans* (1826). He also carries the name of another major character from the earlier story, Uncas, the son of Chingachgook. This link to Leather-stocking's past encourages the trapper to join a quest to rescue the kidnapped woman. Later, Sioux characters abduct Inez as well as Leather-stocking, young lovers Ellen Wade and the bee-hunter Paul Hover, and the Pawnee Hard-Heart, leading to the novel's climactic fight. While the Leather-stocking team emerges triumphant, the emigrants, in fact, give up on their westward quest, after casualties and conflict have proved how ill-suited they are for western life. The novel's non-Leather-stocking main characters eventually marry and return to Kentucky, revisiting the La Platte River region—the scene of the novel's action—one final time to see the dying Leather-stocking and his faithful dog companion, Hector. For Cooper, animals matter; in addition to Hector, a donkey named Asinus is a significant character in the novel. Toward the beginning of the story, the squatters' livestock are stolen by a party of Sioux. Indeed, the theft of these horses and cattle acts a catalyst for several of the events that follow. Even in this novel treatment, Cooper places cows, and their contrasted bison, in the midst of the conflict. While the novel emphasizes the actions and consequential death of Abiram White as the motivation of its plot, the animals play significant roles as well.[20] Throughout *The Prairie*, domestic cattle are shown to be unsuited to the grazing options of the Great Plains, where the bison thrive, and they are one of the facilitators for the conflict between the Euro-American characters and the Sioux. I argue that Cooper uses cows, and to a lesser extent other livestock, in *The Prairie* to show the unsustainability of a Euro-American form of westward agri-expansion. Indeed, as Leather-stocking notes, "The one follows man, the other natur'," privileging the natural.[21]

Cooper provides an example of a wagon train that follows the U.S. migrant route, with the able-bodied walking alongside the livestock, and household goods and those unable to walk traveling within the wagon. Most importantly, the inclusion of domestic animals in the emigrant party makes such an effort possible. Cooper writes: "The vehicles, loaded with household goods and implements of husbandry, the few straggling sheep and cattle that were herded in the rear, and the rugged appearance and careless mien of the sturdy men who loitered at the sides of the lingering teams, united to announce a band of emigrants seeking for the Eldorado of the West."[22] Cattle, horses, sheep, pigs, chickens, and other domestic animals are what establish civilization to families like the Bushes, as well as a sense of nationhood, though they are attempting to move into a new and foreign territory. Instead of a kind of equalizing element, though, the squatters are determined to maintain strict divisions between people and animals. Ishmael Bush, in fact, declares that he does not want to be "treated like one of your meanest cattle!"[23] For a dishonest, harsh man attempting to take advantage of opportunities available in the newly colonized U.S. West, he feels it is important to maintain strict lines between gender, class, race, and even species. As he is the villain character, this attitude is one that readers find inappropriate.

Nature, though, is the great equalizer, especially in a novel set in the Great Plains during the time of the Lewis and Clark Expedition. At this time, the Great Plains are not suited for domestic cattle; rather, this is the domain of the bison. The cows traveling with the squatters are on a seemingly perpetual quest for "good browse": "The meagre herbage of the Prairie promised nothing, in favor of a hard and unyielding soil, over which the wheels of the vehicles rattled as lightly as if they travelled on a beaten road; neither wagons nor beasts making any deeper impression, than to mark that bruised and withered grass, which the cattle plucked, from time to time, and as often rejected, as food too sour, for even hunger to render palatable."[24] This poor browse is contrasted to the prepared pasturage of the East that would be the preferred food for these domestic cattle, largely provided by deforestation as seen in *The Pioneers* and *Rural Hours*. A fickle species, cows are used to a narrow range of foods, and Euro-Americans had made great efforts to provide them with satisfying fodder throughout their shared history. Cooper is here writing of the frontier agriculturally, seemingly in conflict with whether this westward expansion—the movement of Manifest Destiny— was a good or an evil for the nation. He would believe civilization ultimately to be a good, but not one without evils and guilt. Eventually, this would also involve changing the natural grasses that existed in regions like the Great Plains after the Coopers' time. In the decades following both Coopers' writings, the nation would change the Great Plains from the grasslands that sustained the bison, and the people who depended on the bison for their sustenance, to non-native grasslands of wheat, corn, and soy— cereal grains that were grown largely as feed for livestock.[25]

While horses were perhaps the most iconic animal of the western region at this time—as even the Native American nations and tribes of the region recognized and valued the importance of horses—domestic cattle would grow to be the most significant food animal of the continent. In *The Prairie*, the Sioux find value enough in livestock—specifically in horses, but also sheep and cattle—to steal the squatters' animals. "In the midst of this tumultuous disorder, a rushing sound was heard, similar to that which might be expected to precede the passage of a flight of buffaloes, and then came the flocks and cattle of Ishmael, in one confused and frightened drove."[26] This theft, in fact, is one of the key motivators of the text—it is the event that gets the plot moving and develops the following action. Leather-stocking comments when he realizes what had occurred, "They have robbed the squatter of his beasts!"[27] Ishmael Bush then details what the Sioux have stolen, what he has lost. In addition to eight mares and a foal, he laments, "the woman has not a cloven hoof for her dairy, or her loom," referencing the cattle and sheep through their much-needed products.[28] He is considering the natural world only as it is mediated by its domestic production. The Bush family are already known to be predators and exploitative, as they have kidnapped a woman for ransom before moving west. The loss of their livestock, though, is a further blow, and by this act the Bushes "are changed from farmers into horseless hunters," according to William Vance.[29] Leather-stocking proceeds to help the Bushes regain their livestock from the Sioux. In order to reclaim the property of the squatters, Leather-stocking releases, or steals, the Sioux's horses, creating a distraction and maintaining a kind of balance between the two parties.[30] Horses had been adapted to aid and fit into the lives and cultures of the people who used them during the nineteenth century. They were not specifically a food animal in U.S. culture, however; domestic cattle, on the other hand, primarily were raised for food. For the stability of that food source, Euro-Americans had learned to adapt their natural environments to fit one particular consumption style.[31]

As a contrast, Cooper introduces a bison stampede in the middle of *The Prairie*, one in which the trapper parts the tide of bison like Moses parting the sea. Contrasted to the unfitness of the squatters' cows to the region, the bison are thrilling to witness in their native, natural prairie. "A few enormous Bison bulls were first observed scouring along the most distant roll of the Prairie, and then succeeded long files of single beasts, which in their turns were followed by a dark mass of bodies, until the dun-coloured herbage of the plain was entirely lost, in the deeper hue of their shaggy coats."[32] Cooper compares even the coloring of the animals against the similarly colored landscape, only slightly differing and equally fitting. Moreover, Leather-stocking observes, "There go ten thousand oxen in one drove without keeper or mas-ter, except him, who made them and gave them these open plains for their pasture!"[33] Leather-stocking compares bison to cattle directly, calling the bison "oxen," an often interchangeable word for cattle. Indeed, this is not the only time in the novel Cooper compares domestic cattle and bison. Here, the trapper presents his idea of "natur'" on the prairie, rather than the domestic cat-tle traveling with the squatters and comparatively unfit for this challenging environment. What follows is a scene where Leather-stocking stands with his arms outstretched and seems to part the sea of bison around his party, one that includes men, women, and a donkey. However, it is actually the bray of Obed Bat's don-key, Asinus, that ultimately saves the party from being trampled by the stampeding bison. "All their efforts would have prov'd fruitless however, against the living torrent, had not Asinus, whose domains had just been so rudely entered lifted his voice, in the midst of the uproar."[34] Even the donkey is more fitting in this scene than the humans, including Leather-stocking. Asinus accomplishes what Leather-stocking fails to do; the donkey turns the bison away in the midst of their stampede. Throughout much of the novel, various distinct animals are seen to be comparable, their primary difference being—in addition to size—their ability to survive and thrive in the Great Plains without dramatically damaging the land for humans, animals, and other forms of life.

The bison are used to the land, whereas the domestic cattle—and perhaps the U.S. squatters and those who travel with them—

are unsuited to the new terrain and style of agriculture required. Cooper's descriptions of the representative animal of the Great Plains, the bison, is as influential on U.S. readers here as Irving's descriptions were in his western writings like *A Tour on the Prairies* (1832). Throughout *Cattle Country*, I argue that domestic cattle are the representative U.S. animal. Bison, on the other hand, quickly become associated with the Native communities who rely on them for food and other products, an important part of the culture of the nations and tribes of the Great Plains. Bison are a significant contrastingly representative animal in the United States again in the twenty-first century, with numbers growing in protected spaces like Yellowstone National Park. They have come to symbolize—or perhaps have always symbolized—a representative wildness at the heart of the continent. Interestingly, one representative kind of cattle seems to oppose the other kind; domestic cattle and bison seem unable to exist on the land at the same time. Where one lives, the other disappears, in the nineteenth or twenty-first century. In the early nineteenth century, Cooper seems to be arguing for the unsuitability of Euro-American cattle—and perhaps the proportional appropriateness of the bison—in the Great Plains. Over the course of the nineteenth century, after his writing, bison would be cleared away, of course, to make the Great Plains more suitable for the domestic cattle of U.S. agri-expansion.[35]

One of Cooper's main concerns in *The Prairie* is the idea of ownership and settlement of the new territory—and he leaves it ambiguous as to who truly deserves to live in the Great Plains region—as the Bushes and the rest of their party return east at the end of the novel. The Euro-Americans, then, are not ready for westward expansion in the 1830s, according to Cooper. In his 1950 introduction to *The Prairie*, which he names "Cooper's own favorite," Henry Nash Smith recognizes the importance of the Leather-stocking tales in terms of "the frontier experience on the

American character."[36] I would add that Cooper also connects the bison to the Native American communities of the Great Plains, specifically the Sioux and the Pawnee. The Sioux character Weucha asks, "Have the pale faces eaten their own buffaloes, and taken the skins from all their own beavers . . . that they come to count how many are left among the Pawnees?"[37] A variation of this question is repeated at the end of the novel, as Mahtoree tells Hard-Heart: "Where a Pale-face comes, a red-man cannot stay. The land is too small. They are always hungry."[38] Although these statements are made by villain characters as opposed to the idealized Hard-Heart, they remain valid. The significance of the bison is already present in *The Prairie*, a half-century before they would be marked for extermination by U.S. policy as a form of warfare against Great Plains nations and tribes. In this way, Cooper already questions the western movement by the Euro-Americans into the U.S. West whose true expansionist drive would come later in the nineteenth century.[39]

Cooper's controversial attitudes regarding Native American characters continues in *The Prairie*. He presents two main stereotypical, underdeveloped types of Native American characters: the Sioux villain Mahtoree and the Pawnee hero Hard-Heart. These correspond with similar figures in his other novels, like the idealized Uncas in *The Last of the Mohicans* (1826), for example, as well as his similarly stereotypical Euro-American characters. Leather-stocking early engages with the two communities as types: "The Pawnees and the white men are brothers, but a Sioux dare not show his face in the village of the Loups."[40] By the end of the novel, though, a Pawnee, Hard-Heart, will become the closest thing to a son that a dying Leather-stocking has. Cooper shows that, like the bison, the Native Americans of the Great Plains have a reason for being in the region that the squatters lack, one that is more appropriate to the "natur'" of the region. Although he still treats them as doomed, they are also strong, courageous, moral, and even proud. The fact, indeed, that the Bushes are called "the squatters" shows their lack of an appropriate claim to the land in the first place. This charge is reasserted when they are forced to return east at the conclusion of the novel. However, Leather-stocking—Cooper's voice for proper resource management and bridge figure between Native American and Euro-American culture—likewise has nowhere left to travel, and he dies at the beginning of this expansionist moment.[41]

Cooper wrote *The Prairie* as a largely theoretical statement about his ideas for a growing U.S. frontier during the first half of the nineteenth century. When Leather-stocking lights out to the territories following the conclusion of *The Pioneers*, he is attempting to escape Euro-American "civilization." However, the settlers followed him. The Bushes contrast Leather-stocking's independence and appreciation of "natur'." Indeed, the trapper is Cooper's ideal of a man of "natur'," a term he uses throughout the novel. However, Cooper also privileges "civilization" over the "savage" elements of the Native American nations and tribes of the East and the Great Plains. In *The Prairie* he to a large extent equates cows and other livestock with civilization and Euro-Americans, and bison with Great Plains peoples and savagery. However, he complicates this reading by placing the domestic cattle in the possession of the Bush family, disrespectful squatters who do not and should not remain on the prairie. Moreover, the Sioux and Pawnee characters, while largely stereotypes, do show a variety of characters among Native Americans, beyond a pan–Native American exemplar. Throughout, Cooper maintains that Euro-Americans involved in U.S. agri-expansion should be good stewards of the land. They must learn to appreciate the frontier as it is, rather than force it into their mold of civilization. In her nature writings, Susan Cooper would continue this discussion, perhaps even taking it farther in terms of stewardship and conservation. She also privileges a Christianized, conservative civilization over the manners of the Iroquois and other Native tribes and nations within her experience. However, she worries about an exploitative form of this civilization, one that would lead to agri-expansion, deforestation, and the reduction of species diversity, all of which she analyzes in her 1850 natural history work, *Rural Hours*.[42]

Susan Fenimore Cooper's Natural Rural Cattle

Susan Fenimore Cooper surpassed her father in terms of writing creative nonfiction, publishing her best-known work, *Rural Hours*, in 1850. Editors Rochelle Johnson and Daniel Patterson note that "*Rural Hours* presents what ultimately emerges as Cooper's argument for a sustainable balance between human culture and its natural surroundings."[43] Indeed, Cooper argues for a version of sustainable agriculture as early as the mid-nineteenth century. In *Rural Hours* she and her family visit a local farm, describing their experiences there in detail. Throughout the work she idealizes small-scale farming, warns her readers of the dangers of overharvesting and deforestation, alludes as well to the Iroquois who inhabited the region of Cooperstown before the Euro-American settlers, and finally voices her fears for the West at the hands of the frontier communities. In *Rural Hours*, Cooper places herself in the constellation of nineteenth-century U.S. nature writers, predating Thoreau's *Walden* (1854) by four years. Lawrence Buell names *Rural Hours* "the first and still the most ambitious seasonal compendium published by an American author."[44] Cooper neatly lays out this domestic image in her journal-style natural essay that focuses on the flora and fauna of the Cooperstown area over the course of four seasons, or one year (see fig. 6). I argue that in her writing, Cooper argues for a more stewardship-based form of agriculture to counter the expanding U.S. frontier during the nineteenth century. She performs this work through her detailed observations of natural phenomena and the human-connected nature of agriculture.

Fig. 6. Artistic representations of nature adorned Susan Fenimore Cooper's work, providing another visual element to complement her descriptions of the natural world, as well as her fears for its destruction through resource mismanagement. *Rural Hours* title page (New York: George P. Putnam, 1851). Biodiversity Heritage Library.

As early as English colonization of North America, the cows brought with the Europeans were appreciated for their meat and dairy products. Although pigs and chickens were the primary meat-producing animals during the early period of U.S. history, cows were even then predominant as aspirational representative animals for Euro-American farmers. The nation expanded west in the early nineteenth century largely because of the needs of their livestock. Cooper presents her agricultural ideal against this fact, and her attention to place in Cooperstown is intentional. The small and the local are privileged over the expansive. Descended from William and James Cooper, living in a town named Cooperstown, Susan Cooper felt the weight of legacy in her rootedness to the Otsego Lake region. Indeed, the early Coopers were instrumental in agricultural pursuits in the region, founding early agricultural societies. Susan Cooper's focus on the domestic and agricultural is thus familiar. To this, though, she also added an interest in resource management, of protecting wild spaces from intensifying agriculture. Her nature writings combine an appreciation for the natural world with a rural form of Euro-American agriculture, opposed to developing U.S. agri-expansion. Thus, the key issues of resource management and stewardship had already become a significant concern in her writings.[45]

Cattle and other livestock therefore play a dominant role throughout *Rural Hours*, though not as part of an expansionist agriculture. Rather, Cooper describes her example of the B—— family farm, her ideal U.S. farm, as moderate, not expansive. Her approved form of agriculture is one that manages its resources appropriately, practicing proper stewardship of the land. The elderly couple running this farm seek to maintain themselves rather than to grow more wealth:

They kept four cows; formerly they had had a much larger dairy; but our hostess had counted her threescore and ten, and being the only woman in the house, the dairy-work of four cows, she said, was as much as she could well attend to. One would think so; for she also did all the cooking, baking, washing, ironing, and cleaning for the family, consisting of three persons; besides a share of the sewing, knitting, and spinning. We went into her little buttery; here the bright tin pans were standing full of rich milk; everything was thoroughly scoured, beautifully fresh, and neat. A stone jar of fine yellow butter, whose flavor we knew of old, stood on one side, and several cheeses were in press.[46]

The entirety of this picture is connected to dairy cows—the predominance of cattle, while not beef cattle, is noteworthy in this scene. Even a retired farm woman considers four dairy cows to be a necessity to a well-maintained household. This image of idealized rural domesticity follows the standards of nineteenth-century U.S. domestic management, which would be appreciated by the likes of Catharine Beecher or Lydia Maria Child.[47] Mrs. B——'s dairy products include milk, butter, and different kinds of cheese. Formerly, Cooper observes, the family cared for more cattle in their dairy. In their retirement, though, their desires have changed, presumably, and therefore they have reduced their investments in cattle. This picture runs counter to the expansive ideal of the nineteenth-century U.S. frontier. Reducing is privileged over expanding in this example, and Cooper returns her reader to earlier U.S. agrarianism, idealizing small farms. The B—— family farm can be physically and economically sustained by three people, the dairy portion of the farm run by one elderly woman. Cooper thus encourages her readers to appreciate this kind of life, as opposed to the squatter frontier habits described by her father in *The Prairie*.

Cooper next depicts the agricultural production of goods on the farm, where the B—— family produces almost all they consume. This is a standard that would be challenged—really, placed under threat—following industrialization and agri-expansion. "In the spring, a calf was killed; in the fall, a sheep and a couple of hogs; once in a while, at other seasons, they got a piece of fresh meat from some neighbor who had killed a beef or a mutton. They rarely eat their poultry—the hens were kept chiefly for eggs, and their geese for feathers."[48] This was the assumed normal level of production even into the nineteenth century. People living on these small-scale farms consumed mostly beef, pork, and poultry that they produced on-site. When James Cooper would have been writing, though, the nation was already committed to moving west; therefore, by the time Susan Cooper wrote, a generation later, the idea of this farm was already nostalgic, no longer the national norm. Along with western expansion was agricultural intensification, where cattle and other domestic animals were being produced in greater surplus quantities, no longer at the subsistence level. However, the agricultural practices of this ideal farmstead connect to Susan Cooper's overall argument in *Rural Hours*, as she here acknowledges the need for Euro-Americans to emulate this idealized farmstead, rather than the agri-expansionist model. Agriculture is a noble act; Cooper does not argue for humans to avoid manipulating the natural world. However, she argues that they should practice an appropriate level of respect for the nonhuman world. The B—— family farm becomes Cooper's example of appropriate stewardship of the environment through agriculture.[49]

While writing passionately about conservation, Cooper is decidedly not against agriculture in *Rural Hours*. She applauds the increase of crop production. Although she mostly describes vegetable crops, she also notes the herds of cattle in the area with pride: "Year after year, from the early history of the country, the land has yielded her increase in cheerful abundance . . . and at eventide the patient kine, yielding their nourishing treasure, stand lowing at every door."[50] Cooper indicates her attitudes toward moderate increase, the natural "cheerful abundance" of the "patient kine," or cattle, their beef and dairy a "nourishing treasure," maintaining her national and cultural attitudes about the appropriate uses of land and agriculture. Indeed, she reminds her readers, "It needs not to be a great agricultural establishment with scientific sheds and show dairies . . . a simple body, who goes to enjoy and not to criticise, will find enough to please him about any common farm, provided the goodman be sober and industrious, the housewife be neat and thrifty."[51] Cooper argues against the expanding, modernizing element in U.S. agriculture during this period; rather, she encourages her readers to return to a small-scale form of agrarianism that includes livestock.[52] She does not want the pure form of nature that does not include humans, but rather the "middle landscape" of Leo Marx. Moreover, she emphasizes good stewardship throughout *Rural Hours*, both in her exemplary farm family scene and in her later warnings of deforestation and displacement of the Iroquois, the destruction of bison populations on the Great Plains, and the decimation of human populations that would be occasioned by that loss.

Interestingly, Cooper spends little time on the people who lived in the Otsego region before her now-threatened small-scale Euro-American farmers. The Iroquois are described largely using the "vanished Indian" trope throughout her writings. She notes, for example, that "there are already many parts of this country where an Indian is never seen."[53] She goes on, though, to acknowledge that Cooperstown is physically "within the former bounds of the Six Nations, and a remnant of the great tribes of the Iroquois still linger about their old haunts, and occasionally cross our path," adding significantly, "when it is remembered that the land over which they now wander as strangers, in the midst of an alien race, was so lately their own—the heritage of their fathers—it is impossible to behold them without a feeling of peculiar interest."[54] Cooper does try to value the fact that the land whose change she currently laments is being lamented by another community as well. This doubled lamentation, though, is easy for her neighbors of Cooperstown to neglect, as they are able to ignore the Iroquois if they desire. The Iroquois appear only "occasionally," and their presence "haunts" rather than becoming a tangible experience. Importantly, though, the Iroquois are not gone, she admits, and indeed are "interesting." Paula Gunn Allen warns against taking a static view of Native American characters in literature: "Stasis is not characteristic of the American Indians' view of things. As any American Indian knows, all of life is living—that is, dynamic and aware."[55] While lamenting the history of Euro-American and Native American relations in the Otsego region, Cooper does not offer suggestions that would allow the nations and tribes of the area to practice or return to their own agriculture. Her ethic of stewardship responds to the loss of small-scale farms to those of the expanding U.S. West, as well as to a lesser extent the earlier losses felt by the Iroquois and other Native American tribes and nations.[56]

As part of her challenge to the attitudes of the nation during westward migration and agri-expansion, Cooper imagines the results of the Indian removals and Indian Wars that would last much of the rest of the century by describing her fears for the bison of the Great Plains: "The reckless extermination of the game in the United States would seem, indeed, without a precedent in the history of the world. Probably the buffaloes will be entirely swept from the prairies, once covered with their herds, by this generation."[57] Indeed, this is an allusion to James Cooper's discussion of the bison in *The Prairie* and the incursion of the domestic cattle into that space with the squatter family, the Bushes. In keeping with her ongoing argument for sustainable domestic agriculture, rather than the expansionist tendencies she was witnessing and feared would lead to environmental degradation (even if she would not have used that language), Cooper also laments the fate of the Iroquois and reads that onto the peoples of the Great Plains. In some ways, her romanticization of the B—— farm is a negation of Iroquois domestic history, a negation that has dated back through at least three generations of Coopers. In addition, though, Cooper's discussion of her fears of deforestation and the extermination of the bison can be compared to scenes in James Cooper's novels like the passenger pigeons in *The Pioneers* and even the bison stampede in *The Prairie*, especially as she was her father's amanuensis. In some ways, then, Susan Cooper uses this conflation to foreshadow the bloodshed to come for the Great Plains nations and tribes as well as the animals living in the path of westward expansion, namely the bison.[58] Three entities are compared in Susan Cooper's literalized discussion of natural occurrences in Cooperstown: the local flora and fauna, the Euro-American farmers, and the indigenous farmers. While not going so far as to privilege the agriculture of the Iroquois over that of her Euro-American neighbors, she credits them for not destroying their natural habitat in the way that she fears Cooperstown farmers—not including the B—— family—were doing through deforestation and overharvesting. Although she is not suggesting dramatic changes to U.S. agricultural development, she decidedly prefers small-scale farming over the large-scale model that was becoming a greater part of nineteenth-century U.S. agri-expansion.[59]

Rural Hours is an innately ethical work, in keeping with her family's Christian background. However, as opposed to many of her contemporaries' forms of Christianity, Susan Cooper's ethics involved an appreciation of the natural world for its own sake. While she showed through farms like that of the B—— family that a certain kind of human intervention in nature was appropriate, she also argued that the nation's settlers must exercise care and humility as they continued developing and expanding into the western frontier. In her writing she intends her model farm to be useful for the nation as it moved farther west over the course of the nineteenth century, focusing particularly on her village's agriculture. Her stewardship philosophy desires people to be able to use agriculture appropriately, but also to make it possible to be used so in the future. She is not a Leather-stocking figure, preaching the ways of nature. Rather, throughout her writings she engages with the "civilization" of the village and encourages others to give more emphasis to agricultural practices that maintain and sustain the land, rather than overuse those resources.[60]

Conclusion: A Small-Farm Call to Action

Susan Cooper returned to the issue of domestic animals in a practical way in her 1869 essay, "Village Improvement Society." There, she argues again for a specific kind of agriculture, one that is both sustainable and domestic. By 1869 the frontier looked vastly different from that seen in James Cooper's Leather-stocking tales or even her own contribution in *Rural Hours*. Following the Civil War, Susan Cooper attempted to bring the nation together through neatly maintained villages, the rural life that was quickly vanishing through expansion, intensive agriculture, urbanization, and industrialization. In this essay she maintains the superiority of the United States over the "Old world": "How different is the state of things to-day, and in our own country! Village life as it exists in America is indeed one of the happiest fruits of modern civilization."[61] Johnson and Patterson name this essay a call to action as they write that the village had become "the most important cultural site in the country," adding that "the future of the nation's moral and aesthetic character depends on improving our villages."[62] In Cooper's writing, though, success is not the conclusion. She argues for the creation of Village Improvement Societies: "The finishing touches for a village on the prairies, or one on the sea-shore, or one in the Green Mountains, in Oregon, or in Texas, should, of course, vary very greatly in some of their details. But the spirit, the intention, must be everywhere the same. . . . [T]o increase, in short, its true civilization, that is to be our aim."[63] Her village is no longer solely Cooperstown. After 1869, such idealized rural villages, with small-scale agriculture like Farmer B——'s, might be found across a number of regions of the expanded United States. She seeks, therefore, to improve U.S. agriculture on a larger scale. Specifically thinking of animal husbandry, she observes, "Of course, your streets should be protected by confining all cattle, pigs, poultry, to the grounds of their owners," addressing the closing of the commons, and foreseeing again the future of the U.S. West, in the closing of the plains following the invention of barbed wire.[64] Her key interests from *Rural Hours*, specifically her attention to focused, small-scale details of the local flora and fauna, appear in clearer, more strident tones in this essay. She has moved beyond the theoretical to the direct.

The Coopers were writing against the increase of industrialization, expansion, and intensification in the nation that would come during the second half of the nineteenth century. Susan Cooper's work is often compared to Thoreau's. Both nature writers worked to change their respective societies, including their agriculture, through the respect for nature. This ideal is a theme that would continue for both writers into their later works. Susan Cooper calls for land stewardship of a kind that would eventually lead not only to Thoreau but also to writers like Aldo Leopold and Annie Dillard in the twentieth century and beyond. Johnson observes that, opposed to his style in *Walden* (1854), Thoreau's Journal grows more like Susan Cooper's style of writing over time: "In his Journal, Thoreau wrote increasingly in an 'objective' style . . . toward a prose that represents these [natural] things as accurately—and as literally—as possible."[65] I would add that Thoreau became more concerned with accurate representations of nature in his later writings as well, specifically *The Maine Woods* (1864) and *Cape Cod* (1865). These texts become the subject of the following chapter, as I show that in his later works Thoreau argues against agri-expansion involving cattle and the intensive production of other domestic animals. Instead, he privileges local cuisine—flora and especially fauna—in natural essays describing moose in Maine and clams on Cape Cod.[66]

THREE

Henry David Thoreau, Regional Cuisine, and Cattle

"Think of the consummate folly of attempting to go away from *here!*" writes Henry David Thoreau in his Journal on November 1, 1858, "when the constant endeavor should be to get nearer and nearer *here.*"[1] A localist sensibility is to be expected in a naturalist like Thoreau, engaged as he was in the study of regional plants and animals. His focus on these as food sources is a significant part of his engagement as a naturalist with flora and fauna. In the mid-nineteenth-century United States, though, Thoreau's interest in regionally specific foods situated him in opposition to the prevailing trends in food culture, which were increasingly toward consumption of nationally cultivated and marketed foods, especially beef and dairy products of the expanding cattle industry. Thoreau's interest in food matters, then, led him to question the wisdom of the movement toward industrializing the nation's food supply, focusing particularly on New England. His travels, into the local woods and swamps of Concord or extending north to Maine and south to Cape Cod, entailed experiments with locally based menus that presented an alternative to the currently popular one that he viewed as ill-founded. He sought a different way to view consumption through his living experiments as well as his travels. The foods he consumed in his travels took on an importance equal to that of the beans and woodchuck discussed in *Walden* (1854).[2] Exploring the Concord and Merrimack Rivers with his brother in 1839, for example, Thoreau was interested in the fish population, especially the shad. Later, on a hunting excursion with his cousin in the Maine woods, he sampled moose and found it palatable. He also recalled that while walking the Cape shore with friend Ellery Channing, he lit a small fire on the beach and roasted and ate a giant clam—only to regret doing so when his stomach rebelled later that evening. (Thoreau comforted himself in remembering that the Pilgrims, arriving in Provincetown Harbor, similarly became sick eating a species of shellfish.)[3] Throughout the nineteenth century, settlers were bringing their domestic animals with them to the frontier territories of the expanding nation. In a period when the United States was engaged in intense self-definition to accompany its dramatic physical change, a time when Ralph Waldo Emerson called for a uniquely U.S. literature, Thoreau considered a corresponding U.S. culinary culture as he addressed indigenous foods that had a special relationship to place, that reflected local bioregions, and that honored Native traditions. Indeed, his "Indian notebooks," a record of his wide readings in anticipation of a projected work on indigenous America, evidences his considerable interest in the food cultures of the various nations and tribes of the Northeast. At a time when the nation was experiencing dramatic expansion, including the development of an intensive livestock husbandry and a national focus on cattle production, Thoreau's journeying—physical and philosophical—was regional: he was drawn to New England's wilderness areas and its "grossest groceries."

The national trend toward meat—specifically beef—bred on the western range continued, however, reflecting the continent's abundant grazing land, especially as Native American communities were driven off these same lands and territories derived from the Louisiana Purchase of 1803, the Indian Removal Act of 1830, and the Treaty of Guadalupe Hidalgo of 1848, all of which greatly increased U.S. areas available for raising livestock and the cereal grains to feed them. Looking for territory on which to replicate a British model, a 1606 emigrant to Virginia—seeking wealth—considered cattle first: "I have seene many great and large Meadowes having excellent good pasture for any Cattle," as noted by Virginia DeJohn Anderson.[4] By the arrival of the *Arabella* to Massachusetts Bay Colony in 1630, the need for emigrants to travel with domestic animals was obvious, as the 700 emigrants from Europe brought 240 cattle and 60 horses with them.[5] Cattle outnumbered horses by a large measure, and they were quickly becoming known to be a lucrative commodity. As early as the first New England settlements, then, colonists marketed cattle to the sugar plantations of the Caribbean, representing an early example of global trade. Euro-Americans turned forests into grasslands, which were then converted by grazing animals like cattle into dairy, meat, and other goods and labor. Indeed, livestock were a key element in the Euro-American settlement of the frontier—including the early frontiers of Massachusetts and Maine, as well as the extending frontiers of the West in nineteenth-century United States.[6]

Thoreau's interest in regional cuisines against a backdrop of a developing U.S. culinary monoculture has been previously documented in the 1985 presidential address of the Thoreau Society Annual Meeting, where Frederick Wagner discussed culinary passages from *A Week on the Concord and Merrimack Rivers* (1849), *Cape Cod* (1865), and *The Maine Woods* (1864). *Cape Cod*, he notes, "struck his sensibilities. The reverberations were in his stomach," while in *The Maine Woods* "the reader perceives a heightened appetite."[7] Expanding upon Wagner, I argue that Thoreau's wild regional dining adventures were a necessary part of his writing specifically because they countered the developing national norm toward far-reaching animal production enterprises—domesticated animals like cattle, pigs, and chickens expanding across the U.S. West to feed the growing industrial centers of the East. Furthermore, Thoreau enjoyed reading travel narratives, learning through his reading that the Hottentots ate the marrow of the kudu raw, for example.[8] Thoreau's literary models described a variety of cultures and traditions that would influence his own experiences—physical and literary. In *A Week on the Concord and Merrimack Rivers*, Thoreau describes one of his primary motivations for travel: "These modern ingenious sciences and arts do not affect me as those more venerable arts of hunting and fishing, and even of husbandry in its primitive and simple form."[9] "Hunting and fishing" would lead directly to ingestion, and these "ancient" and "venerable" ways of "husbandry" appear in his descriptions of food production and consumption throughout his works. While Thoreau considered himself to be part of a world growing steadily more modern through its "ingenious sciences," he was less interested in promoting those developments than in their alternatives.[10]

Meanwhile, New England farmers were being driven by such ingenious science to specialize in certain crops and livestock in order to feed the industrializing urban centers and factory markets of the East. Inland Massachusetts communities grew fewer basic food commodities in greater quantities as demand increased and labor forces changed. Until refrigerated train cars made transportation of beef cattle economical at scale, enabling the development of a western cattle industry, Boston markets received two-thirds of their stock from Massachusetts, followed by New Hampshire, Vermont, and Maine. Thoreau would have been most familiar with one area of specialization, the production of beef cattle in the Connecticut Valley for the larger Boston markets. In *Walden*, Thoreau bemoans the cattle being transported from these scattered farms through Concord to the market in Boston: "And hark! here comes the cattle-train bearing the cattle of a thousand hills, sheepcots, stables, and cow-yards in the air. . . . The air is filled with the bleating of calves and sheep, and the hustling of oxen, as if a pastoral valley were going by."[11]

Thoreau recognized the agricultural struggles of his neighbors and friends, like Concord farmer George Minott, to adapt to U.S. territorial expansion and the connected growth of livestock production, and he incorporated these struggles into his writing.[12]

Thoreau was almost uniquely qualified to write about the early industrializing of U.S. agriculture. He witnessed the ending of the subsistence economy of early U.S. history and the beginning of industrialization in his adulthood.[13] Laura Dassow Walls observes that his shortened life span was "long enough to witness and record the arrival of the Anthropocene epoch in America."[14] One key element of the Anthropocene involved the development of the modern diet that began with industrialization. Thoreau's emphasis on regional culinary experiences challenged the Anthropocene that he witnessed in its beginnings, and in his experiments with local flavors he had some of the most positive experiences in his writings. Walls argues that Thoreau's response to the changes he witnessed was optimism rather than despair. On the one hand, he described the decimation of wild animal populations such as wolves, deer, and moose. However, he also celebrated the wildness that remained. The woodchuck described in *Walden* was echoed in the poisonous wild clam in *Cape Cod*. The moose tracks he followed in *The Maine Woods* were being followed by cattle and other domestic animals—and vice versa; the nation's cattle were being left to wander the Maine woods alongside the regional moose.

Identity, including national identity, is defined by culture; Roland Barthes names food "the functional unit of a system of communication" and argues that people in a shared culture—like the United States—communicate through food as much as through other forms of cultural expression, like literature.[15] The study of food is a dynamic method of communicating culture in shared stories, as Mikhail Bakhtin notes: "There is an ancient tie between feast and the spoken word," a connection that may trace to "the very cradle of human language."[16] People identify with food at the same time as—if not before—other cultural markers. As literature works to standardize certain cultural traits and develop new ones, it engages with the discourse of culinary practices as they are being formed. In nineteenth-century U.S. literature, the consumption of food paralleled the consumption of greater expanses of territory. Thoreau responded to this cultural shift through his efforts to consume locally during his travels in Maine and Cape Cod. In his writing he interrogates developing U.S. tastes in his descriptions of eating moose and quahog as part of a deep experience of place at a time when cattle are becoming more dominant fixtures in the national landscape and on the U.S. dinner table.[17]

A Week on the Concord and Merrimack Rivers, Thoreau's first published work and an ode to his brother, John, his traveling companion and best friend who died from tetanus in 1842, is also Thoreau's first investigation into regional cuisine. In particular, Thoreau makes fish an extended metaphor for contemporary U.S. culinary culture. "Whether we live by the sea-side, or by the lakes and rivers, or on the prairie," he writes, "it concerns us to attend to the nature of fishes, since they are not phenomena confined to certain localities only, but forms and phases of the life in nature universally dispersed."[18] As opposed to the other regional foods he would celebrate in Maine and Cape Cod, fish are almost universally available. "It is curious, also, to reflect how they make one family, from the largest to the smallest. The least minnow, that lies on the ice as bait for pickerel, looks like a huge sea-fish cast up on the shore."[19] From cod to minnow to pickerel, fish contain almost the same fundamental shape, purpose, and being. Where there are fish, for Thoreau, there is life. Indeed, John McPhee recognizes that Thoreau could "all but converse" with fish: "he would ripple the water with his fingers, and perch would come to his side."[20]

After presenting general themes concerning the availability of fish, Thoreau localizes his discussion to the Concord region, remarking on the relative poverty of fish species there, as opposed to the dozens of species that would have existed in previous times. Concord, located near Boston and not a wild space of the kind Thoreau usually looked for in his excursions, had already been affected by industry at this time. Thoreau observes that the "Salmon, Shad, and Alewives, were formerly abundant here, and taken in weirs by the Indians, who taught this method to the whites . . . until the dam, and afterward the canal at Billerica, and the factories at Lowell, put an end to their migrations hitherward."[21] As soon as the damming of rivers began, the fish populations plummeted—from overharvesting, but especially from habitat destruction. Thoreau challenges the idea of progress in his claim that damming the rivers as part of the industrialization of New England had devastating effects on the fish of the area. The fish species no longer abundant in Thoreau's Concord hold a special significance for him in the purview of *A Week on the Concord and Merrimack Rivers* as he reads their fate into the industrializing world of his present.

Ultimately, the American shad, which had been so abundant in the rivers of colonial New England, becomes Thoreau's representative fish. Shad are, indeed, one of the nation's "founding fish." Similar to the salmon of the Yukon, American shad are ocean fish that return upriver to spawn; thus, the influx of dams along the major rivers of New England in the first half of the nineteenth century proved disastrous to the species. Thoreau appreciates the dangers to the shad through the destruction of their spawning grounds. He observes, glumly, that the shad are still to be found on the river, though inevitably under duress. They are dammed up, and "the Corporation" seems to have won the battle: "Poor shad! where is thy redress? When Nature gave thee instinct, gave she thee the heart to bear thy fate? . . . Armed with no sword, no electric shock, but mere Shad, armed only with innocence and a just cause, with tender dumb mouth only forward, and scales easy to be detached. I for one am with thee, and who knows what may avail a crow-bar against that Billerica dam?"[22] Thoreau envisions a dramatic solution to the destruction of the founding fish and of his fishing lifestyle along the river: a crowbar applied to the dam. While of course a fantasy, this idea reflects his sense of commonality with the natural world, even with creatures that he considers to be potential food, demonstrating his larger view of the interconnectedness of life. He will occasionally eat a shad, but he wishes fervently that he could protect this same shad from a tragic fate at the hands of those who destroy it in large numbers without the justification of consumption.

This theme of tragedy, as well as anger, resonates throughout *A Week on the Concord and Merrimack Rivers.* Beyond anthropomorphizing the shad, Thoreau equates these fish with his brother: both are transient, patient, and needful of a champion. Additionally, Thoreau believes we should approach nature with gratitude to the life-forms around us. As he theorizes his relationship with the shad in his writing and the consequences of industrialization and the intensification of agriculture that would result in expanding livestock production across the Great Plains region, Thoreau becomes one of the nation's early conservationists, like iconic hunters Daniel Boone—a model for James Fenimore Cooper's Leather-stocking character—and President Theodore Roosevelt.[23] This is not a characterization often applied to Thoreau. While he is well known as an environmentalist and his work has become synonymous with wilderness conservation, he has not often been linked to the hunters who were some of the earliest concerned parties in terms of habitat depletion and large-scale destruction of animal populations. Indeed, Thoreau is more often recognized as a preeminent early vegetarian.[24] While he was often of two minds regarding hunting, he fished and did go on hunting parties as part of his larger experience of the wildness he valued. In the "Higher Laws" chapter of *Walden*, his strongest argument for moral consumption, Thoreau observes:

"Fishermen, hunters, woodchoppers, and others, spending their lives in the fields and woods, in a peculiar sense a part of Nature themselves, are often in a more favorable mood for observing her, in the intervals of their pursuits, than philosophers or poets even, who approach her with expectation. She is not afraid to exhibit herself to them."[25] Thoreau's adventures in Maine were most often as part of a hunting party, his distant cousin George Thatcher being an avid hunter. In Maine, when Thoreau's party shoots and butchers a moose, he does not take part in the shooting and is frankly offended by the experience, but he does eat some of the meat. While the shad is his illustrative animal in *A Week on the Concord and Merrimack Rivers*, the moose plays a comparable role in *The Maine Woods*. Thoreau's experience with the moose hunt and with eating moose becomes one of his core subjects in his Maine writings.[26]

Eating Moose in *The Maine Woods*

Thoreau's first expedition to the Maine woods, described in "Ktaadn," is with a party of four men—including his cousin George Thatcher—who are interested in lumber and peripherally in local flora and fauna. The party never encounters moose during this excursion, but the animal's presence is felt throughout. They see moose horns, the footprints of a cow moose and her calf, and at one evening's camp "there was the skeleton of a moose here, whose bones some Indian hunters had picked on this very spot."[27] In this instance, an assumption is made that the moose bones are the result of Penobscot hunters, potentially romanticizing the wildness presented here through a conflation of animal and Native American cultures within the woods during this, Thoreau's first experience in Maine. While his first essay deals largely with climbing Mount Katahdin, he describes other natural elements of the region as well: "The tracks of moose, more or less recent, to speak literally covered every square rod on the sides of the mountain; and these animals are probably more numerous there now than ever before, being driven into this wilderness from all sides by the settlements."[28] Thoreau's theme of spaces and places in harsh contrast appears already in this statement about inter-actions between the moose and humans. Tracks of moose are everywhere, but moose are noticeably absent. There is a process ongoing in which they are becoming more densely packed within a receding wilderness brought about through encroaching U.S. loggers and settlers, as well as their corresponding livestock. Thoreau observes that moose "are so wary, that the unskillful hunter might range the forest a long time before he could get sight of one."[29] Unlike in Concord, where he takes pride in his deep knowledge of the natural world around him, Thoreau is at a loss to see a wily animal like the moose in its native terri-tory: "We remembered also that possibly moose were silently watching us from the distant coves."[30] The moose function as a counterpoint to Thoreau's description of the sublime on Mount Katahdin. Where the mountain is beyond humanity, encountered but unknowable, the hidden moose are directly engaged with people, knowable yet unseen.

In the case of the cattle arriving in Maine with U.S. loggers and settlers, they are both knowable and seen. Thoreau realizes that domesticated cattle are in direct opposition to the regional specificity that moose embody. Cattle are not adapted to the nat-ural conditions of Maine, and Maine has not yet been cleared to accommodate cattle grazing. Describing the essence of Maine, Thoreau observes that "on either hand, and beyond, was a wholly uninhabited wilderness, stretching to Canada. Neither horse, nor cow, nor vehicle of any kind, had ever passed over this ground. The cattle, and the few bulky articles which the loggers use, being got up in the winter on the ice, and down again before it breaks up."[31] At this point, cattle are used mostly in the winter to work with the lumber industry, linked almost exclusively to U.S.

trade. They must remain in Maine only for the winter, when the waterways are frozen solid. With the change of seasons, when the lakes and rivers become impassable, cattle and other domestic animals are supposed to be removed from the region.

Some settlers, however, have started to bring cattle to Maine permanently, a trend that would intensify over the course of Thoreau's visits. He writes of a settler who "kept horses, cows, oxen, and sheep. I think he said that he was the first to bring a plough and a cow so far."[32] Domestic animals have at this point begun to accompany permanent settlement to Maine—as they were starting to do as well across the continent. No longer merely the domain of the Penobscot, lumber interests, and wild animals, Maine was becoming, during Thoreau's time, part of the domesticated United States, the Euro-American cultivated world he had been trying to escape. Thoreau describes a scene where cattle from the lumber industry are required to stay behind in Maine after an early thaw: "I do not know whether Smith has yet got a poet to look after the cattle, which, on account of the early breaking up of the ice, are compelled to summer in the woods, but I would suggest this office to such of my acquaintances as love to write verses and go a-gunning."[33] The cattle are left to themselves and forced—or allowed—to forage among the woods over the summer season, perhaps becoming moose-like or, in fact, evidence of U.S. population expansion into the farthest reaches of the continent. In this transitional phase, Maine pasturelands are not yet a viable large-scale option. Thus the rancher must also play the hunter, and Thoreau implies that the rustling heard in the woods may be from moose or cattle, as poetry and hunting have come together in a satirical fashion. The moose-like cattle in *The Maine Woods* are an incongruity, but one that points to the agricultural model for the nation. They, like the expansion in which they play a significant role, are inappropriate to Thoreau's ideal of indigenous places and their cuisines, represented here by moose.

Thoreau notes that what might seem to be an exotic commodity in Boston or Concord is ordinary fare for the local population of Maine: moose meat is "common in Bangor market."[34] Once again, while Thoreau does not expressly view a moose during this first visit, he witnesses their presence, in this case at the market. Ironically, though, Thoreau's party relies largely on wild trout for their own consumption rather than the store-bought moose meat. He returns to the theme of moose as Maine's representative animal in his second essay, "Chesuncook." During this second trip, Thoreau's party does shoot and butcher a moose. Thoreau considers an essential part of his excursions to be the discovery and sampling of indigenous foods that have a special relationship to place, that weave the eater into the local environment, and that honor Native traditions. Their Penobscot guide, Joe Attean, and the party's consumption of the moose meat following that hunt together becomes one of Thoreau's core subjects in "Chesuncook."[35]

In many ways, Thoreau returned to Maine to eat a moose. His experience with moose hunts emerges as an important theme in his second essay, where he describes another venture with Thatcher and their Penobscot guide, one that reveals the material level of hunting, processing, and consuming a moose. Natives of the northernmost regions of North America, Europe, and Asia, the moose inhabits some of the least-populated spaces of the Northern Hemisphere, and Thoreau considers them the epitome of Maine's wildness (see fig. 7). The largest of the deer family, a mature bull moose can reach sixteen hundred pounds—females closer to twelve hundred—and stand up to seven feet tall, according to Victor Van Ballenberghe.[36] The eastern moose of Maine is one of four North American subspecies. They have distinctive long legs and nose and a square body adapted to the wet northern climate. Thoreau appreciates the relevance of the moose to its native land, at the same time using the nineteenth-century trope of equating wilderness spaces with indigenous cultures: "Such

Fig. 7. Thoreau likely saw a moose and calf not unlike these shortly before they were shot and killed by his hunting party. Library of Congress, Prints & Photographs Division, Carol M. Highsmith Archive, 2015.

is the home of the moose, the bear, the caribou, the wolf, the beaver, and the Indian."[37] Perhaps more than any other animal in his excursions, the moose for Thoreau becomes symbolic of its home territory. William Williamson declares as much in his history of Maine, writing that the moose is honored as a state symbol, like the white pine.[38] The elements of the moose that make it particularly suited to Maine prove its significance to Thoreau during his travels there.

Upon first seeing the cow moose with her calf, Thoreau is reminded "of great frightened rabbits, with their long ears and half inquisitive half frightened looks." Rather than a sublime, intimidating creature, like a bear or cougar, these animals are compared to an iconic prey animal. The world of the moose is not threatening, but rather in its own way threatened, perhaps something to be defended rather than defeated. Declaring them "the true denizens of the forest," Thoreau is shocked by the terrible reality that the hunting of such creatures entails. "I looked on, and a tragical business it was; to see that still warm and palpitating body pierced with a knife, to see the warm milk stream from the rent udder, and the ghastly naked red carcass appearing from within its seemly robe."[39] Thoreau focuses on the udder full of milk, the nourishment this cow moose will no longer be able to provide another, now abandoned, living being. He does not dispassionately describe a natural process in this scene. He expresses some of the same depth of emotion seen previously when he lamented the destruction of the shad spawning grounds of Concord in *A Week on the Concord and Merrimack Rivers*. In Maine, the moose continues to haunt Thoreau as the party leaves the scene of her death: "We could see the red carcass of the moose lying in Pine Stream when nearly half a mile off."[40] Indeed, before the actual killing of the moose by Thoreau's party, Attean reminds Thoreau of the grisly business involved in hunting. Thoreau writes: "We saw a pair of moose-horns on the shore, and I asked Joe if a moose had shed them; but he said there was a head attached to them." In Thoreau's Journal, Attean adds a phrase that in the finished essay Thoreau gives to himself: "They did not shed their heads more than once in their lives."[41] Thoreau records this moose, the discarded remnants of bodies that are not being used to their fullest capacity, before and after his own moose-hunting experience.

An important part of Thoreau's relationship with his Penobscot guides was their deep knowledge of the Maine region. He earnestly sought to learn from Attean, and later Polis, as seen in his focus on their preparations of local foods, specifically moose meat. Their keen awareness of one particular region stood in contrast to contemporary U.S. monoculturalism. Robert Richardson considers the hunting and butchering of the moose in

"Chesuncook" to be "the parallel emotional center" of the experience of "primitive nature" on the mountain in "Ktaadn." The "thrilling sight" of the hunt was of great interest to him once "the moose's blood was shed"; this interest, though, turned into "outright revulsion" once the butchery began—in what Philip F. Gura refers to as an example of "true New England fastidiousness and a bit of humanitarian outrage." Eventually, though, Thoreau's interest in the wilderness experience allowed him to overcome his aversion to hunting itself in order to study the hunting methods of Attean.[42]

When he finally has the opportunity to taste fresh moose meat, what he had previously researched through his reading and his learning from Attean could now be described from personal experience. His interest in studying Maine went beyond the hunting experience and toward an engagement with moose meat as part of a local menu. He consumed his way into the Maine ecosystem: "We had moose meat fried for supper. It tasted like tender beef, with perhaps more flavor; sometimes like veal."[43] In his description of this moose supper, Thoreau's remarking the meat's tenderness and flavor suggests his sensual experience of eating foods native to Maine. Thoreau takes advantage of his opportunity to eat the meat of a newly hunted animal. He also begins to make note of his meals of moose meat: "We here cooked the tongue of the moose for supper,—having left the nose, which is esteemed the choicest part, at Chesuncook, boiling, it being a good deal of trouble to prepare it."[44] Soon, though, he moves toward more developed dishes in which the moose meat is combined with other local ingredients. His epicurean skills in the context of Maine cuisine are becoming more developed: "We also stewed our tree-cranberries. . . . This sauce was very grateful to us who had been confined to hard bread, pork, and moose-meat," and, occasionally, to beef. He can sauce his meat with other local foods—in this case the local tree cranberry, *Vibur-*

num trilobum, which his companions declare to be "equal to the common cranberry."[45] Concentrating now on a variety of local foods, ones bioregionally specific, Thoreau discovers alternatives to the fare lately become standard on the U.S. table.

While he develops more sophistication in his recipes, he also grows more like the narrator of *Walden*, who bases his dietary decisions on "higher laws." During the moose feast at Chesuncook, he leaves a communal dinner "at which apple-sauce was the greatest luxury to me, but our moose-meat was oftenest called for by the lumberers," and goes for a lone walk where "a large slice of the Chesuncook woods" becomes his dessert for the evening.[46] He considers meat, as well as the locally available fish, a necessary travel food, but he comes to prefer the edible plants of the area— like his favorite berries. In this way, Thoreau resembles his *Walden* persona, who desires a diet of simple, mostly vegetable foods. In fact, he physically distances himself from those people who have not lost their taste for meat, moving from literal to figurative eating with the aesthetic consumable of the Chesuncook woods.

Thoreau's desire to understand everything about the moose as the symbolic animal of the Maine woods—and thereby to deepen his understanding of Maine and its indigenous cultures— continues into his interactions with his Penobscot companions. At the end of "Chesuncook," camping with several Wabanaki hunters, he describes a scene in which he has asked his companions for English translations of various Abenaki place-names. They teach him the meanings behind words like *Chesuncook, Caucomgomoc, Pammadumcook, Kenduskeag, Mattawamkeag,* and *Penobscot*, among others. They spend a disproportionate amount of this time analyzing the naming of Moosehead Lake. If all the other names receive a paragraph's worth of discussion, Moosehead Lake, or *Seboomook*, earns a page and a half of analysis. Ultimately, "Tahmunt thought that the whites called it Moosehead Lake, because Mount Kineo, which commands it, is shaped like a moose's head, and that Moose River was so called 'because the mountain points right across the lake to its mouth.'"[47] Tahmunt's focus is on the U.S. story about naming places, not the Abenaki word. At the same time, the preeminence of the moose is significant for Thoreau thematically as part of his larger narrative thanks in part to the animal's importance to indigenous communities. The moose lends its name, along with its size, shape, and style, to such natural elements as a river, rock, and certain kinds of men, in addition to a berry, bird, fly, and tree. Its influence is felt throughout the region. "Moose" appears in the appendix to *The Maine Woods*, where Thoreau mentions key Abenaki words. The moose is far more characteristic of the Maine woods than domestic animals like cattle, in terms of place-names as well as Thoreau's literary focus. In order to understand the region, therefore, Thoreau knows that he must work to understand this animal.

He continues to use the moose as a conversational curiosity, this time in the home of Governor Neptune, senior figure of the tribe. Thoreau describes an interesting element of Penobscot culture using a story of theirs about moose: "[Neptune] could remember when the moose were much larger; that they did not use to be in the woods, but came out of the water, as all deer did. Moose was whale once. Away down Merrimack way, a whale came ashore in a shallow bay. Sea went out and left him, and he came up on land a moose."[48] Fishing and whaling were important elements of the culture and economy of the Northeast leading up to Thoreau's time, and this legend provides a textual connection between the moose of Maine and other iconic large mammals. As the whale dominates the ocean, so the moose has control over the woods of Maine. Thoreau follows this legend with quotes from nineteenth-century scientific treatises about the moose that agree with this legend in some respects—as in the fact that moose might have been larger in the past. Cattle, historically unknown in the region, complete the triumvirate of iconic mammals establishing themselves there. Cattle were developing into the dominant large land mammal in terms of U.S. territory and its agriculture through the nineteenth century, but Thoreau's attention here is toward the moose.

The encroachment of cattle on moose territory was significant by 1853. While Thoreau originally notes in "Ktaadn" that cattle were purely seasonal creatures in Maine, following the loggers and connected entirely to the logging trade, by the time he wrote "Chesuncook" they had started to represent long-term Euro-American settlement of the continent. After the moose hunt, still within sight of the corpse of the killed moose, Thoreau notes pastureland interspersed within the Maine woods: "There was considerable native grass; and even a few cattle—whose movements we heard, though we did not see them, mistaking them at first for moose—were pastured there."[49] In his appendix of "Flowers and Shrubs," Thoreau also connects non-native plants to the cattle who carried them "as far into the woods as Chesuncook," even though he still believes that these same cattle "cannot be summered in the woods."[50] He observes the intrusion of domesticated plants, as well as the animals that conveyed them into the wilderness. During this visit, Thoreau notes more domesticated, livestock-appropriate landscape in Maine. He comments, "We seem to think that the earth must go through the ordeal of sheep-pasturage before it is habitable by man."[51] Indeed, Thoreau could have added cattle as another invasive domestic animal exemplar. His quest to experience Maine wildness through its cuisine occurs as the signs of U.S. expansion become more apparent. This theme of understanding all aspects of the moose contrasted to the rise of a domestic livestock presence in Maine continues in his third essay, "The Allegash and East Branch," as well.

Thoreau is not the only moose-meat gourmand on his last trip to Maine, described in his final essay in *The Maine Woods*. On this visit, he spends less time describing the actual moose and instead provides additional examples of moose-meat dishes. Both his companion Ed Hoar and their Penobscot guide, Joe Polis, contribute to Thoreau's growing knowledge. Polis instructs Thoreau and Hoar on his own methods: "use 'em best wood. Always use hard wood to cook moose-meat." Thoreau continues, describing Polis's process, "He cast the moose's lip into the fire, to burn the hair off, and then rolled it up with the meat to carry along. Observing that we were sitting down to breakfast without any pork, he said, with a very grave look, 'Me want some fat,' so he was told that he might have as much as he would fry."[52] Like any seasoned cook, Polis has particular opinions about the correct way to prepare the foods he knows best. In this case, he roasts the meat and then fries it in grease. Later, Hoar decides to prepare moose meat in his own preferred way: "My companion cooked some in California fashion, winding a long string of the meat round a stick and slowly turning it in his hand before the fire. It was very good."[53] This rotisserie manner of preparing the moose meat adds another dimension to the culinary experience of *The Maine Woods*, a technique from yet another region. The party can bring together and compare regional cuisines reflecting the dramatic geographical changes of the nation during the nineteenth century. While loggers, gold seekers, and other settlers—with their requisite cattle and other livestock—had intruded on the formerly unfenced lands of the U.S. frontier, driving out what and whoever was previously on those lands, elements of specific regions survived through cuisine. Thoreau does not have as intimate knowledge of Californian cuisine as Hoar does, but he is able to appreciate the multiregional novelty of the dish nonetheless.

Thoreau continues to experiment with his deepening appreciation for regional cuisine as he adds fish to the soaking moose tongue to make a kind of stew: "I caught two or three large red chivin (*Leuciscus pulchellus*) early this morning, within twenty feet of the camp, which, added to the moose-tongue, that had been left in the kettle boiling over night, and to our other stores, made a sumptuous breakfast."[54] Thoreau's descriptive adjectives highlight his growing pleasure in eating the wild foods of Maine. The moose tongue and chivin breakfast is nothing less than "sumptuous." He often uses contradictory images in his writings, and in this case the idea of a sumptuous breakfast is removed from a luxurious indoor setting and located rather at a humble Maine campfire. Thoreau is the chef, experimenting with previously unfamiliar native plants and animals in their own context. The moose becomes more flavorful when augmented by other wild Maine foods—berries and fish, for example. Thoreau's simple intention is clear as he writes, "I wished to learn all I could before I got out of the woods."[55]

Thoreau furthermore references a comment made by Thatcher that gets to the essence of the cuisine of Maine's wilderness in contrast to U.S. expansion and consolidation of cuisine—plentiful and cheap meat focusing on cattle, along with pigs, sheep, and chickens. After a dinner of fried moose, Thoreau writes of Thatcher, "when hunting up the Caucomgomoc about 2 years ago, he found himself dining one day on moose-meat, mud turtle, trout, and beaver, and he thought that there were few places in the world where these dishes could easily be brought together on one table."[56] Meanwhile, though cattle were still mostly seasonal animals in Maine, they were leaving more evidence of their presence, increasingly remaining in the region beyond the logging season. He observes an island that "was not inhabited, had only been used as a pasture for cattle which summered in these woods."[57] Thoreau's earlier comment that cattle do not summer in Maine has now been countered by more recent evidence. Maine offers a variety of foods that would have been unavailable elsewhere in the nation and that at least suggest the possibility of a different paradigm of food consumption from the emergent industrializing one.[58]

While "The Allegash and East Branch" is most often analyzed for Thoreau's relationship with Joe Polis, the party's Penobscot guide, there is much to be learned from his study of local foods as well. Thoreau's concept of a traveler is one who exists largely in Leo Marx's "middle ground," even among the rivers of Maine. In addition, Thoreau works to translate his experience with the moose into text, and he turns this experience into a theme throughout *The Maine Woods*, especially in the second and third essays. The moose, furthermore, contrasts the encroaching cattle. Conceptually, wildness is often contrasted to civilization in Thoreau's work, and in that vein, moose can also be contrasted to cattle—as similarly representative animals.[59]

The moose would remain an illustrative animal for Thoreau for the rest of his life. According to biographer Walter Harding, among Thoreau's final words were "Indians" and "moose."[60] The concept of studying—even to the point of consuming—moose in the Maine woods resonated with him up to his death. Thoreau's writings about eating moose meat in Maine are significant to his project of establishing an idea of regional food and literary culture based on knowledge of venerable places. Returning to "Chesuncook," Thoreau applies this theory to his own experience as part of a moose-hunting party:

> I think that I could spend a year in the woods fishing and hunting just enough to sustain myself, with satisfaction. This would be next to living like a philosopher on the fruits of the earth which you had raised, which also attracts me. But this hunting of the moose merely for the satisfaction of killing him—not even for the sake of his hide, without making any extraordinary exertion or running any risk yourself, is too much like going out by night to some woodside pasture and shooting your neighbor's horses. These are God's own horses.[61]

Thoreau's emerging environmental ethic develops through experiences with this kind of bioregional subsistence. He does not, in the end, approve of moose hunting in Maine, seeing it, ultimately, as too domesticated. Comparing this form of hunting to "shooting your neighbor's horses," Thoreau brings the nation's large-scale land allotments, stocked with domestic animals, negatively to the fore. Domestic horses, like cattle, represented the encroachment of U.S. expansion into increasingly distant sections of the continent. The wild element of Maine is reduced to another example of U.S. settlement, and moose are compared directly with horses and, by extension, other domestic animals associated with that settlement. As Thoreau laments this development of U.S. culture, his culinary adventure becomes a practice that resists a particularized national ideology. Rather, in *The Maine Woods* he argues for a reverence for a specific area carefully rooted in local culture, history, and rituals of subsistence. The variety of local foods available in a region like Maine becomes a prize. Thoreau carries this level of detail to his final excursion-based writing, in his descriptions of preparing and consuming shellfish in, along with the encroachment of cattle to, *Cape Cod*.

Clams and Cattle in *Cape Cod*

Thoreau visited Cape Cod four times between 1849 and 1857. During his first visit, in 1849—the same year as the publication of *A Week on the Concord and Merrimack Rivers*—he is delayed and detoured from his original itinerary by a shipwreck in which ninety-two people died, mostly Irish immigrants.[62] In his first chapter, "The Shipwreck," Thoreau speculates on the vicissitudes of nature, and by the later chapter "The Sea and the Desert" he comes to the realization that "naked Nature" is "inhumanly sincere, wasting no thought on man."[63] This awareness colors the rest of his first excursion, the route of which provides the plot and direction of *Cape Cod*. While Thoreau's posthumous *Cape Cod* is often humorous, it is also perhaps his darkest work.[64] Playing on his theme of regional cuisines, Thoreau describes his culinary exploits on Cape Cod as instances of humor, misunderstanding, and even of danger. Negative experiences involving local foods are rare in Thoreau's travel works, but he does record, as mentioned above, an unfortunate experience eating a giant quahog, a significant food item for the original inhabitants, the Nausets, but one requiring appropriate culinary knowledge for its preparation, knowledge that Thoreau as a visitor lacks.[65]

Clams have a particular significance for Thoreau, as they relate directly to the Native populations of coastal New England who used clams as food and currency. The Wampanoags and Nausets, for example, used them for their wampum, and Mark Kurlansky observes that the Lenape of the New York region "regarded hard clams as money, especially the large ones with bluish inner shells that New Englanders call quahogs."[66] Thoreau describes the omnipresence of shellfish on the Cape: "I saw that their shells, opened by the Indians, were strewn all over the Cape. Indeed, the Cape was at first thickly settled by Indians on account of the abundance of these and other fish. . . . [O]ysters, clams, cockles, and other shells, mingled with ashes and the bones of deer and other quadrupeds."[67] For Thoreau, this evidence of the Nausets is key to his interest in Cape Cod. While there was a time in which shellfish were considered waste foods by Euro-Americans, they made a sumptuous cuisine that had maintained human populations on the Cape for thousands of years. Thoreau focuses on his contemporaries as well as the Nausets, describing a local wrecker—one who makes a living scavenging from the area's shipwrecks: "He looked as if he sometimes saw a dough-nut, but never descended to comfort; too grave to laugh, too tough to cry; as indifferent as a clam—like a sea-clam with hat on and legs, that was out walking the strand."[68] Lacking in features, a shellfish makes an unlikely metaphor for a human; however, Thoreau draws a striking and fitting parallel between clams and a certain kind of local living near the sea. Clams possess several traits that translate to human features, like toughness and tight lips that could represent gravity or indifference. Thoreau's way of describing humans as clam-like, and potentially clams as possessed of free will or malevolence, will be significant again when he attempts to eat a quahog.[69]

On the third day of this Cape Cod visit, Thoreau describes preparing and eating a particular species of quahog for the first time, following the traditional regional fare of the "clam bake" (see fig. 8). He assumes he will be able to cook this giant clam much as he would a clam found closer to his native Concord, as a local wrecker assures him that this "was the sea-clam, or hen, and was good to eat."[70] The wrecker does not warn Thoreau about the piece that must be removed before preparation. Either that man assumes Thoreau already knows, he is playing a trick on the ignorant tourist, or perhaps he simply forgets to provide that detail. In any case, Thoreau lacks a vital piece of knowledge regarding his meal.

> We took our nooning under a sand-hill, covered with beach-grass, in a dreary little hollow, on the top of the bank, while it alternately rained and shined. There, having reduced some damp drift-wood, which I had picked up on the shore, to shavings with my knife, I kindled a fire with a match and some paper, and cooked my clam on the embers for my dinner; for breakfast was commonly the only meal which I took in a house on this excursion. When the clam was done one valve held the meat and the other the liquor. Though it was very tough, I found it sweet and savory, and ate *the whole* with a relish. Indeed, with the addition of a cracker or two, it would have been a bountiful dinner. I noticed that the shells were such as I had seen in the sugar kit at home. Tied to a stick, they formerly made the Indian's hoe hereabouts.[71]

Thoreau goes into some detail regarding this meal that he and Channing have found at their feet—one of his most comprehensive dining scenes in *Cape Cod*. He describes the kind of

Fig. 8. *A Clam-Bake*, 1873. Winslow Homer's classic sketch of a clam bake provides a version of what would become one of Thoreau's most infamous experiences with regional foods. Smithsonian American Art Museum, The Ray Austrian Collection, gift of Beatrice L. Austrian, Caryl A. Austrian, and James A. Austrian.

wood used, the setting, and even the plating of the clam. He pronounces the clam to be "very tough" but asserts that he likes it anyway, because of, not in spite of, this element. This seeming inconsistency resonates with his earlier descriptions of wild foods. Finally, he jokes of the element "of a cracker or two" being all that is lacking from providing him a "bountiful" meal—like the "sumptuous" Maine breakfast of moose meat and chivin stew. So far, Thoreau's meal looks like his other experiments with wild foods on his excursions. However, his lack of appropriate regional culinary knowledge, and his lack of a local guide, will return to harm him. Thoreau foreshadows this later distress by italicizing "the whole"; he eats the entire clam, not removing the key piece that renders his meal poisonous. His italics become an inside joke for his readers who understand how to properly cook quahog.

After establishing this scene, Thoreau continues with its inevitable denouement. That evening, he stays with the family of John Young Newcomb, a retired oyster fisher. Newcomb informs him that while "the great clams were good to eat," people should always be careful to remove "a certain part which was poisonous, before they cooked them," as that item "would kill a cat" upon ingestion. Hearing this information, Thoreau smugly believes that he must be "tougher than a cat," as he was not feeling any ill effects.[72] However, the local knowledge proves true, and later that evening Thoreau starts "to feel the potency of the clam which I had eaten." He is "obliged to confess to our host that I was no tougher than the cat he told of." Even at that time, Newcomb is gruff with the traumatized Thoreau, saying that his malaise "was all imagination." Regardless, Thoreau is by now feeling the effects of food poisoning; he writes, "I was made quite sick by it for a short time, while he [Newcomb] laughed at my expense."[73] In this instance Thoreau is able to laugh along with Newcomb, a bit of levity against himself. His painful lesson is a good one, and he passes on the knowledge of his experience to his readers. Thoreau seeks regionally specific foods, but there are risks involved in eating unfamiliar foods without guidance, and he must be more careful next time. By the end of his quahog-eating adventure, Thoreau has come to realize that he is no stronger than the proverbial cat and that the local oysterman has more correct information than the proud narrator-tourist.

Moreover, the clam itself is connected to his earlier shipwreck scene. Thoreau's particular giant clam, the *Mactra solidissima*, or *Spisula solidissima*, had only come to him because it had been deposited on the shore from the sea bottom during a storm. "Our host [Newcomb] told us that the sea-clam, or hen, was not easily obtained; it was raked up, but never on the Atlantic side, only cast ashore there in small quantities in storms."[74] The same storm that wrecked the *St. John* might also have deposited this clam at Thoreau's feet. Thoreau chooses this dramatically obtained, essentially wild, and ultimately dangerous bivalve for his experiment in wild foods. Although he fails to make a delicious, healthy meal from this clam, he does make a valiant attempt to access all the facts before cooking it. His intercalated themes of misinformation, shipwreck, and bodily distress make his clambake a different experience in wild foods than those he relates in *The Maine Woods* or *A Week on the Concord and Merrimack Rivers*. In the other excursions, Thoreau has either deep knowledge of regional foods or a trusted guide to introduce him to new methods of preparing them. In *Cape Cod*, Thoreau literally bites off too much on his own. His possible Nauset guides no longer live in the area, and the clam-like U.S. locals keep to themselves until it is too late for him. He will continue to play on themes of incongruity and food—of eating wrongly—in his later chapters, where he describes the cattle he witnesses on the Cape.

Thoreau's final chapters of *Cape Cod* engage significantly with the role of cattle on the people and the natural environment of the Cape. Walking Cape Cod, Thoreau is reminded constantly about its alien landscape. Returning to themes he established in his first chapters, he comments in a later chapter about the sublimity of the sea from the Cape: "It is a wild, rank place, and there is no flattery in it."[75] However, he also notes repeated signs of its domestication. Old women milk cows, and herds of cows mingle with windmills in the sandy vista. Much of Thoreau's project in writing is to reconcile these disparate views of Cape Cod. He reads Cape Cod as a balance between wildness and appropriateness. Ultimately, Thoreau comes to realize that domestic animals are not natural elements of this wild place, but rather an incongruous signal of Euro-Americans and their livestock taming the wilderness during the expansionist nineteenth century.[76]

As he stays at the lighthouse on the Cape, Thoreau comments on his nautical surroundings: "The air was so moist that we rarely wished to drink, though we could at all times taste the salt on our lips. Salt was rarely used at table, and our host told us that his cattle invariably refused it when it was offered them, they got so much with their grass and at every breath."[77] Noting that cattle eat this "beach-grass," Thoreau spends some time studying how it is able to grow in sand.[78] This form of grass becomes another key element of the Cape Cod environment for Thoreau during his expedition. It is specific to the area—unlike grasses found elsewhere in the United States, or even the New England region. Indeed, this is the same beach-grass on which Thoreau sat when eating the improperly prepared quahog. He reads Cape Cod as a moored boat, with the beach-grass as its connecting cables: "Thus Cape Cod is anchored to the heavens, as it were, by a myriad little cables of beach-grass, and, if they should fail, would become a total wreck, and erelong to go the bottom. Formerly, the cows were permitted to go at large, and they ate many strands of the cable by which the Cape is moored, and wellnigh set it adrift, as the bull did the boat which was moored with a grass rope."[79] This boat is under attack by the cattle. He extrapolates a dramatic possibility from a familiar example. Readers could understand the issue of a bull unmooring a small craft by eating hempen rope. Thoreau develops that image into an allegory of the real threat of overgrazing on the Cape. Cows eat grass; that is their nature. However, Cape Cod is not a natural prairie, and it is unable to withstand this grazing over a long period of time. Thoreau worries that these domestic animals might "sink" the Cape itself.[80]

By Thoreau's time, the locals have had to alter their animal husbandry practices in the light of the threat to the Cape grasses: "They therefore recommended that beach-grass be set out on a curving line over a space ten rods wide and four and a half miles long, and that cattle, horses, and sheep be prohibited from going abroad, and the inhabitants from cutting the brush."[81] Cattle are no longer allowed to graze freely on the Cape Cod commons.

They are now herded, fenced, and in other manners treated more as they would be among the developing ranch lands of the U.S. West. Cape Cod is becoming more like other U.S. regions, even though it is unsuited for this kind of domestic animal production and agriculture. It is not a prairie, and treating it like other kinds of grasslands will not succeed in the long run. As the clam that poisoned Thoreau is not a native clam but was dumped on Cape Cod out of the Atlantic, so Euro-Americans and their domestic animals have been as unceremoniously dropped upon the Cape to the detriment of Cape Cod itself.[82]

Thoreau continues describing the seemingly grotesque eating habits of the local cows—which, lacking proper grass, are discovered to eat the area's fish: "It is rumored that in the fall the cows here are sometimes fed on cod's heads! The godlike part of the cod, which, like the human head, is curiously and wonderfully made, forsooth has but little less brain in it,—coming to such an end! to be craunched by cows! I felt my own skull crack from sympathy."[83] Cows should not be eating fish, and the fact that they are disturbs him enough that he wonders, "What if the heads of men were to be cut off to feed the cows of a superior order of beings who inhabit the islands in the ether? Away goes your fine brain, the house of thought and instinct, to swell the cud of a ruminant animal!"[84] Even after being reassured by some locals that the story he has overheard is truly a fish tale, Thoreau remains unsettled. Once again, he has to question his basic belief system in terms of nature and what survives in a location like Cape Cod. In his manner of comparing himself to that fish being consumed, as well, Thoreau draws attention to his own inappropriate culinary behavior on the Cape. Ruminating on the incongruity of the cow's diet, he continues to demonstrate his interest in anthropomorphizing animals. Thoreau chooses the iconic cod—then as abundant in the North Atlantic as the shad had recently been in New England rivers—as the exemplar fish in this story, and, with cattle particularly in mind, he warns against forcing a homogenized form of agriculture and cuisine upon an unsuited, vulnerable region.

Historical and regional cuisines are as important to Thoreau as the people who lived on Cape Cod. The Nausets are not interviewed in *Cape Cod*, but they are described throughout the text via historical documents. Thoreau establishes early on the interactions between the British and the Nausets. Citing the Nausets' conviction that "there was not any who owned" the Cape, Thoreau notes with irony the British, who became "Not Any's representatives" and claimed the land as their own.[85] He details one of the earliest examples of the conflicts between Euro-Americans and indigenous peoples—the theft of land that would occur repeatedly throughout the nineteenth century and beyond. Indeed, Thoreau's experiences over his four visits to Cape Cod would be the inspiration for his lifelong study of Native American cultures, his "Indian Books" where he amassed almost three thousand pages of research on their cultures and histories.[86] In addition to his interest in Cape Cod's original inhabitants, Thoreau finds himself fascinated by the current locals. Gura notes: "The people whom he encountered on the Cape, the Welfleet Oysterman, the keeper of the Highland Light, the men and boys who sailed from Provincetown harbor in the mackerel fleet, all could read the immense landscape of the region, a literacy that Thoreau came to admire and, eventually, to emulate."[87] Thoreau appreciates the locals, who are by turns salted and preserved in a way he realizes is beyond him in his life at Concord. This interest in researching cultures—specifically northeastern Native American cultures, but also Euro-American regional ones—would result in his visits to the Maine woods and the travel narratives stemming from both sets of excursions.

Thoreau describes his high regard for nature and diet throughout *Cape Cod*; however, he is alternately amused and disgusted by the foods consumed regularly and safely by residents of the

Cape.[88] These clam-like people are able to eat foods that nearly bested Thoreau when he made the attempt. Failure, sickness, and even death are also natural processes—ones that Thoreau witnessed throughout his life and during his Cape Cod excursions. During these visits, Thoreau was preoccupied with ideas of health, given his own poor health—the tuberculosis that ran through his family and would ultimately kill him—and the other diseases that affected his friends and relatives. Meanwhile, in his introduction to *Cape Cod*, Robert Pinsky describes Thoreau's writing as a forerunner of modern travel narratives, adding that it "remains the place's best portrait."[89] The themes of natural simplicity as a way to maintain strength and health that Thoreau enumerates in *Walden* also appear in this posthumous work. Although the storm that precedes the opening scene of the narrative is the cause of great suffering, Thoreau returns to his sense of the natural as ideal. He is able to make his peace with his understanding of "naked Nature," his material discomfort, and his overall well-being through his engagement with local cuisine and culture in *Cape Cod*.[90]

Conclusion: Regional Cuisine versus Cattle

Thoreau exhorts his readers to make an effort to take an expansive view of history when considering regional cultures and their cuisines as workable alternatives to domestic farm animals like cattle, pigs, and chickens. While the Cape looks "frivolous" to readers, *"viewed from the shore,* our pursuits in the country appear not a whit less frivolous" to nature and other humans, Thoreau argues.[91] Perspective is key, he maintains. In addition, the deep knowledge connected to region is something that must be maintained. Thoreau feared a United States defined by sameness, including a sameness of consumables. Pinsky notes: "Part of Thoreau's genius is that he understood modern American life as it was first forming—almost before it formed."[92] Walls agrees, perceiving that Thoreau is not a precursor to globalization and the industrialization of U.S. food systems; rather, he is at the forefront of this industrialization, critiquing it as he watched it gain force.[93] In *A Week on the Concord and Merrimack Rivers*, Thoreau presents what might be his overall thesis on regional cuisine and U.S. culture in the nineteenth century: "There should always be some flowering and maturing of the fruits of nature in the cooking process. Some simple dishes recommend themselves to our imaginations as well as palates."[94] Thoreau's travel narratives give him the opportunity to meditate on life's necessities, connecting food, morality, and the imagination. This transcendental image—that we, like the foods we eat, are warmed, grow, and blossom through our consumption, returns the reader to Thoreau's overall optimism in terms of dietary challenges during U.S. expansion of the nineteenth century. While the nation was becoming the land of cheap and readily available meat, Thoreau sought to slow his food process down, to simplify it. His wish to slow the pace of the nation's expansionist diet, though, is not fulfilled. Rather, agri-expansion continues at a frenetic pace throughout the mid- to late nineteenth century. In the following chapter, Sarah Winnemucca Hopkins must defend her people, the Paiutes, from westward expansion and its consequences as it reached beyond the Great Plains to the Great Basin of the West. She and her people faced removals, destruction of food and land sovereignty, and dehumanizing treatment and death. In her writings, she too uses cattle as the dominant symbol for U.S. expansion. She also compares the desire of U.S. settlers for land on which to raise cattle with the treatment of the land's original inhabitants.

FOUR

Cattle and Sovereignty in the Work
of Sarah Winnemucca Hopkins

In 1867 I was interpreter for my people; but even then they
had nothing. The game has all been killed, except a few rabbits. The pine trees have all been destroyed, so that we can get
no more nuts. The cattle have trampled out the grass in our
little valleys, and we can dig no more roots. If the white people leave us, to go over the mountains to California, as some
people tell us, we must go over the mountains with them too,
or else starve. If we cannot get wild game, we must take tame
game, like cows or steers; the same as the white people would
do if they had nothing to eat, and nothing to feed their wives
and little ones with.

—SARAH WINNEMUCCA HOPKINS, "The Pah-Utes"

The interconnected themes of food and justice emerge in the
writings of Sarah Winnemucca Hopkins through her rhetorical
use of cattle. In "The Pah-Utes," published in *The Californian*
in September 1882, Winnemucca shows her people, the Paiutes, starving due to the destruction of their established harvesting and hunting grounds during U.S. expansion and the
late-nineteenth-century cattle boom.[1] She describes cattle as
harming the Paiutes in specific ways. They have "trampled out

the grass," destroying a necessary Paiute food system, causing a real threat of starvation; Winnemucca explicitly stresses the lack of food available to her people, using terms like "nothing" (three times), "no more" (twice), and, most dramatically, "starve." In fact, the Paiutes have been driven to occasionally stealing cattle and other "tame game"—hunting domestic animals on private lands when wild game and open lands are no longer available to them. Winnemucca describes some of the history between Paiutes and U.S. cattle ranchers leading up to the theft of "cows and steers." This issue in part would lead to the tragic Indian Wars of the nineteenth century, including the Bannock War of 1878, in which Winnemucca herself played a major part.[2] As this epigraph shows, Winnemucca engages with cattle as a significant material symbol in key writings on Paiute food sovereignty.

In her work, Winnemucca seeks to expose the corrupt dealings of reservation agents and achieve some measure of food justice for a community forced beyond the brink of starvation by repeated removals and unfair land policies, highlighting a series of cattle-related food injustices. Janet Fiskio may be discussing Winnemucca as she states: "While the emergence of food studies as an academic discipline has brought new attention to food in indigenous life and literature, work for food sovereignty in indigenous communities precedes this movement by centuries."[3] As a significant part of U.S. expansion in the nineteenth century, cattle were directly or indirectly responsible for much Paiute hardship. Like other nineteenth-century authors, therefore, Winnemucca used a number of rhetorical practices to show that those who may be regarded as subjected and assimilated peoples were in truth employing existing discursive methods for their own causes. These concerns permeate her most famous work, her self-narrative, *Life among the Piutes: Their Wrongs and Claims* (1883); her newspaper articles; as well as her 1884 congressional testimony. As seen in the epigraph, Winnemucca analyzes the changes for the Paiutes before and during westward expansion in her writing. I argue that she uses cattle deliberately in her self-narrative, articles, and testimony to gain sympathy from white audiences, including U.S. government officials. Throughout, she describes the settlers' actions against her people in terms of cattle in key ways: she explains the impact of cattle on Paiute subsistence before and after colonization; she employs her skill with narrative to show how U.S. settlers used the need for ranchlands as justification for Paiute removal; and she stylistically appropriates the idea of livestock to symbolize the condition of the Paiutes themselves.

Cattle are necessary symbols for understanding the food sovereignty portion of Winnemucca's rhetorical project: writing, performance, and testimony. In the nineteenth century, traditional foodways of the Paiutes were diminished through U.S. landgrab policies. Paiutes were driven to a commodity economy, forced to learn U.S. systems of agriculture, including ranching, and eventually driven to famine conditions on a series of reservations. Food sovereignty was of critical concern to Winnemucca during this time. Moreover, in September 2007 the United Nations published the "Declaration on the Rights of Indigenous Peoples," a document designed to encompass the struggles of indigenous communities across the globe, focusing on food as a primary concern. Perhaps Article 31 of the declaration makes the clearest statement about Winnemucca's concerns throughout her work—the importance of native place to a people and the deep knowledge of food and other resources within a specific region: "Indigenous peoples have the right to maintain, control, protect and develop their cultural heritage, traditional knowledge and traditional cultural expressions, as well as the manifestations of their sciences, technologies and cultures, including human and genetic resources, seeds, medicines, knowledge of the properties of fauna and flora, oral traditions, literatures, designs, sports and traditional games and visual and performing arts."[4] The concept of traditional knowledge in culture, specifically food culture, pertains most to Winnemucca's life and work. In her time, the Paiutes had a deep "knowledge of the properties of fauna and flora" of the Great Basin region of North America. However, they were forced, violently and with many casualties, into other territories. Since then, their struggle, one that Winnemucca embraced, had become an effort to recover some of what they had lost. Gary Paul Nabhan gives a specific example of the loss of traditional food knowledge following colonization and expansion: "The Owens Valley Paiute had a field day collecting them [*Pandanus* moth larvae] off desert plants to eat. But such harvests are few and far between these days."[5] What Nabhan and the United Nations describe in the modern period is a continuation of an issue that started with colonization and was exacerbated by westward expansion in the second half of the nineteenth century, as cattle ranches encroached on Native American foodways. For Winnemucca, narrative offered a primary way to improve Paiute sovereignty, specifically in terms of the intersection of food and culture at this early stage of globalization, industrial food production, and corrupt governance.

Winnemucca chose the symbol of cattle in her work for its strong rhetorical value. One of the earliest domesticated animals, raised primarily as a commodity for its meat and dairy, cattle came to the Americas with European colonists. I argue that what was a significant relationship in Europe became if anything more so in the United States—where technological and political changes allowed cattle to be raised across greater distances and in larger numbers than ever before. The establishment of populations of cattle and other domesticated animals was a key project for early colonists as well as with Native American communities as they sought to navigate their changing world. Meanwhile, Spanish cattle became especially significant in what would become the

U.S. West, where cattle ranches dominated agriculturally, even before the arrival of U.S. settlers and their British cows. Cattle were inexorably linked to Euro-American expansion from the period of colonialism through the end of the nineteenth century and beyond.[6]

The quest for new and fresh grazing lands for livestock was a primary motivation for settlement throughout U.S. history. The United States had established itself as the land of cheap and plentiful meat as early as the seventeenth century, the time of John Winthrop in New England, when livestock took on a commodity status. Cattle and other livestock animals were no longer used solely for subsistence, but rather had become part of a global economy entwined with regions like the Caribbean. As New Englanders turned cheap grazing lands into profit through beef cattle, Caribbean colonizers focused on increasing their profits in the production of sugar—perhaps the largest global industry of the sixteenth and seventeenth centuries. In the nineteenth century that industrial spirit became, if anything, more intense. The United States expanded its agricultural economic interests across the continent. This activity led directly in the 1880s to a cattle boom across almost the entire western United States, and Winnemucca's work and adult life coincided with the height of this "cattle fever."[7]

Therefore, Winnemucca used cattle as a representative animal in her quest for Paiute sovereignty as a form of the genre of food studies in literature, which lies at the intersection of race and environmental studies. Jeffrey Myers associates the work of Winnemucca and other writers of color with nature writers of the same period, providing examples including the Progressive Era formation of Sierra Club in 1892 and the National Association for the Advancement of Colored People in 1909. He connects "race and ecology as interrelated themes of American literature from the late eighteenth century to the early twentieth."[8] People of color and environmentalists of the period worked to fight against forces seeking to subjugate both under one dominant society. Meanwhile, a counternarrative emerged in which much U.S. conservation work restricted Native American subsistence and was used to justify Indian removal. Native American communities were seen as misunderstanding the true purpose of nature; they actively engaged with natural spaces that the U.S. government had determined should be left untouched. Furthermore, Paul Outka discusses the natural environment and race in terms of Euro-American and Native American communities in particular: "This association between human subjects and a commodified nature existed in sharp contradistinction to the other central intersection between white supremacy and environmental practice: the association between Native Americans and the precapitalist 'untouched' savage wilderness."[9] Interestingly, Winnemucca uses conservation arguments in her effort to protect Paiute sovereignty. She writes against the move toward the commodification of nature, animals, and even of her own people while navigating the stereotype of Native Americans as "savage" figures in an "untouched" territory in her lived experience of the "cowboys and Indians" narrative.[10]

In "The Pah-Utes," Winnemucca describes traditional methods of acquiring food during her youth, for her a time of relatively cheap and plentiful food, before the influx of both U.S. settlers and their cattle. As cattle become a predominant theme for her later life in her writings, other plants and animals dominate her earlier memories, before the era of abusive U.S. resource allotment policies. She describes the foods that the Paiutes produced to support themselves for thousands of years in the Great Basin.

Our men used to hunt, and after that, our women go into the valleys to gather different kinds of seeds. The men go to fish along the Humboldt and Truckee rivers. They dry game of all kinds, and lay it up for the Winter. Later in the fall men hunt rabbits. . . . Then [in the spring] they go to the fishing-grounds; and when the roots begin to grow, the women dig them up. The name of this root in Indian is called *yah-bah*, and tastes like carrots. They boil them, like potatoes, and use them in soups, and also dry them. Another root is called *camas* root—a little root that looks like chestnuts; and *kous* root, which tastes a little like hard bread. In early days, when white people came among us, they used to eat our food, and compare it with theirs. The same toil was gone through with every year, to lay up the winter supplies; and in these days they always seemed to have plenty of food, and plenty of furs to keep them warm in the winter time.[11]

The Paiutes historically have had "plenty of food" and "plenty of furs," and Winnemucca consciously repeats the word "plenty" in her essay. This is a play on the idea of a "land of plenty" or even the promise of cheap and plentiful meat that the United States had claimed for itself, as well as biblical allusions to plenty readily recognizable to her audience.[12] Rather than beef or dairy from cattle, the Paiutes here eat fish, rabbit, and other game. Moreover, the words *yah-bah*, *camas*, and *kous* are not even translated into English; these foods would have been almost entirely unfamiliar to readers. Interestingly, William Clark does mention such foods in his journal based on his famous 1804 expedition with Meriwether Lewis. Preceding Winnemucca by more than fifty years, he notes: "J. Collins presented us with verry good *beer* made of the *Pa-shi-co-quar-mash* [camass] bread."[13] Winnemucca demonstrates her proficiency in both cultures by offering U.S. consumer-ready foodstuffs—potatoes, chestnuts, and hard bread—as recognizable comparators to the still unfamiliar root vegetables. She provides a rhetorical level of expertise by claiming knowledge of the early U.S. settlers who described these roots—the same roots that were being destroyed by the overgrazing of domestic cattle—in such a manner. Through anecdotal evidence, settlers are shown to support her claims about traditional foods. Therefore, she implies logically that in supporting her knowledge of Paiute cuisine, settlers might support her other, larger claims for greater Paiute food sovereignty. She represents what any Paiute would have known but what every settler taking over this region for livestock ranches either lacked knowledge of entirely or knew but ignored out of disinterest. The arrival of U.S. settlers and their cattle devastated the lives of the Paiutes, and Winnemucca's work shows her trying to help her people survive these cattle-based changes. Winnemucca's *yah-bah*, *camas*, and *kous* roots exemplify the traditions of indigenous Paiute lands, and in her rhetorical work Winnemucca contrasts them to the cattle connected to Euro-American settlement.

Winnemucca's biography shows the foundation for her life's activism. She was the daughter of Old Winnemucca and the granddaughter of Truckee of the Numa, named the Northern Paiutes by Euro-Americans, in what is now the state of Nevada. The Numa bands had lived in the Great Basin area for up to four thousand years: "from the Owens Valley in California through most of western Nevada and into southern Oregon and Idaho," according to Sally Zanjani.[14] Winnemucca, named Thocmetony, or Shell Flower, was from a family of distinction among her band. Like her father and grandfather, she learned early to work with U.S. Americans, as seen in a series of live performances with her family across California and Nevada in the style of Buffalo Bill's "Wild West Show," "a series of 'Tableaux Vivants Illustrative of Indian Life,'" as described by A. LaVonne Brown Ruoff. These live acts grew into a lecture series on the plight of the Paiutes, first presented in San Francisco in 1879. Winnemucca's lectures were successful, and she next presented them on a tour of the East, delivering more than three hundred talks in all. During that time she became friends with activists Elizabeth Peabody and Mary Peabody Mann, who would help her publish *Life among the Piutes*. She also traveled to Washington DC, where she met (and was promptly forgotten by) President Rutherford B. Hayes and spoke to Congress on behalf of the Paiutes. Throughout, Winnemucca proved her skills as a performer, lecturer, and writer on the topic of Paiute sovereignty, continually using the metaphor of cattle.[15]

Intriguingly, Winnemucca was more directly engaged with horses than cattle in her early life. Growing up, she would listen to her grandfather, Truckee, tell about Euro-Americans with abundant food and animals like horses. Indeed, Winnemucca famously rode horseback for hundreds of miles as scout and interpreter for the U.S. Army during the Bannock War. Meanwhile, cattle were also an obviously substantial part of Paiute history. Winnemucca's childhood was a period of enormous change. Even before Indian removals officially began, the Paiutes had few choices in maintaining traditional foodways. Their limited options seemed to be under threat by encroaching cattle, which brought diseases like typhus into the water supply of the Humboldt River region. While horses were seen as a benefit to the Paiutes, cattle were considered a danger even during this early period—competition for scarce resources and a sign of Euro-American westward expansion.[16]

In "The Pah-Utes," Winnemucca follows her discussion of traditional Paiute foods with a paragraph connecting land, cattle, and war. She frames the Bannock War in terms of cattle and scarcity: "The white man wanted [the land], because the roots were good for his cattle, and could make milk and beef and hides and tallow; so he tried to rob them of these lands. They did not like this, and because he despised them, and would give them no redress, they killed him. But the cattle alone were not the cause of this war. The agents were worse than the cattle: what the cattle left the agents took."[17] Winnemucca's language no longer describes a plentiful land. Now, the indigenous people and the land are "rob[bed]," "left," "despised," and without "redress." Following this action, Winnemucca describes a series of consequences that leads to escalated tensions, death, and eventually a brief but tragic conflict. In her description of the causes of the Bannock War, cattle are the motivating factor for U.S. ranchers and their hunger for land. In this passage, she compares humans to cattle in a manner that she returns to throughout her writing. If Paiutes are treated like cattle in other works, now the reservation agents are "worse than the cattle" in terms of the cost to the Paiutes. In either case, the Paiutes have lost their lands, and cattle have been largely responsible for this loss. Winnemucca shows the devastation brought about by cattle ranches on former Paiute lands through her portrayal of one indigenous community facing the encroachment of U.S. settlers and their livestock. Therefore, when Winnemucca describes cattle as a food source and symbol more than horses, pigs, or chickens, she makes a conscious rhetorical choice, one that can be studied in terms of narrative and food sovereignty.[18]

Throughout her life, Winnemucca compares U.S. settlers, their cattle and other livestock, and the Paiutes. She uses a variety of rhetorical strategies in her work as a way to support the Paiutes and combat U.S. territorial expansion, industrial agriculture, and the commodification of the environment. Malea Powell argues: "Winnemucca's text *is* a rhetorical performance, one in which she represents herself as a participant in the kinds of Indian-ness that would have appealed to her late-nineteenth-century reformist audience, that is, the 'civilized Indian' and/or the 'Indian Princess.'"[19] Carolyn Sorisio adds that Winnemucca incorporates elements from both cultures into her public persona, creating a hybrid figure that plays on the artificial concept of the "Indian princess" while critiquing it, making her a form of "postindian warrior."[20] Indeed, Winnemucca's maneuvering of U.S. expectations of what determined authenticity is one of her rhetorical strengths; she maintained a favorable relationship with the press, even during a difficult period of U.S.–Native American relations. In this way, Winnemucca's work shows her rhetorical skills toward her goal of Paiute food sovereignty.

Cattle as Scapegoats for the Agents

U.S. settlers used cattle as justification for the appropriation of Paiute lands and resources throughout the second half of the nineteenth century. In her various narratives, Winnemucca discusses her concerns with the treatment of her people by land agents and ranchers in which the Paiutes suffer under U.S. policies. In an 1870 letter to Indian Affairs superintendent of Nevada, Henry Douglas, Winnemucca laments U.S. policies against the Paiutes, an indictment powerful enough to be included in the appendix of Helen Hunt Jackson's *Century of Dishonor* (1881): "If we had stayed there, it would be only to starve. . . . If this is the kind of civilization awaiting us on the reserves, God grant that we may never be compelled to go on one, as it is much preferable to live in the mountains and drag out an existence in our native manner."[21] While "dragging out" an existence can hardly be considered ideal, it is clearly preferable in this case to the "civilization" of the reservations resulting from U.S. expansion and the encroaching cattle ranches on Paiute lands. Away from their native territory, forced onto a series of reservations, the Paiutes have lost everything. Discussing a "politics of famine" from a Canadian historical perspective, Fiskio's theory of "leftovers" also occurs in Winnemucca's writing on cattle ranchers acting against the Paiutes in the United States. Paiutes, U.S. settlers, and land agents are linked through cattle—as a food source, subject of land-distribution policies, and symbol of the United States in the West. Winnemucca's rhetorical use of cattle is critical to a complete understanding of her work. In this section I focus on her self-narrative, articles, and testimony to show that her writing and performances are some of the most powerful sites where Winnemucca uses cattle symbolically as part of her argument for Paiute sovereignty as ranchers and land agents manipulate laws to benefit themselves at the expense of the Paiutes.[22]

In addition to her focus on unfair resource allotment, Winnemucca traces the expansion of cattle into the region in terms of economic and political control. U.S. ranchers altered whatever territory they entered—whether forest or prairie—to make conditions appropriate for raising cattle and other domestic animals. Barbed-wire fences went up across formerly communal lands.[23] Predator animals were seen as threats and killed en masse. Settlers privatized formerly wild lands into ranches, which were stocked primarily with cattle, pigs, chickens, and horses. In this case, "livestock" is almost always synonymous with cattle, and the Euro-Americans destroyed the lands Paiutes depended on for their traditional way of life through cattle ranches as much as through the direct burning of stored food supplies, deforestation, or overhunting. Later, of course, the entire Great Plains (also referred to as the Great American Desert) would be transformed into the breadbasket of the world, which was really the breadbasket of the many cattle being raised on ranchlands across the U.S. West, in order to be slaughtered in midwestern cities like Chicago and Cincinnati. Lands became privatized, and eventually one agricultural system transformed and dominated the other. The expansion of cattle territory corresponded with the decimation of the U.S. bison, as well, as President Ulysses S. Grant's administration in 1876 acted to "destroy the commissary" of the nations and tribes of the Great Plains through the destruction of the wild bison population.[24]

The scale against which Winnemucca and her people were positioned is exemplified by an archetypal cattle ranch of the period. In the mid-1880s, the cattle ranch of a northern Californian named Henry Miller occupied 1.4 million acres, across

California, Oregon, and Nevada. The Miller & Lux cattle-ranching business dominated the West, providing an example of the sheer size of the cattle ranches during the cattle boom period. Such ranches could encompass entire foodways, making them inaccessible to the indigenous communities who had relied on them for thousands of years. These pernicious land policies influenced labor practices, as well. Paiutes removed from their lands were forced to learn to work as laborers on cattle ranches—often forcibly on reservation lands—that were either wholly unfamiliar to them or previously their own hunting and fishing grounds. This was a continuation to some extent of the western mission system of the Franciscan priests, as we will see in the following chapter. Furthermore, a photograph from 1927 shows a "shipment of 50 cars of cattle from White Pine County, Nevada." Although this photograph was taken much later than Winnemucca's writing, it represents what she would have been fighting, across her native region, throughout her adult life (see fig. 9). In part because of this agro-economical monopolization, Winnemucca fought for sovereignty against settlers who transformed Paiute lands into cattle ranches, forcibly moved the Paiutes onto unfamiliar reservations, and then refused to let them work those lands or gain employment by other agricultural means.[25]

In *Life among the Piutes*, Winnemucca connects cattle directly to the U.S. government, a powerful symbol of westward expansion and the loss of Paiute food sovereignty. Cattle are shown to be a major form of competition for the Paiutes, who had to struggle with a domestic animal for resources. The land agents and ranchers clearly win this conflict.

> Now, dear readers, this is the way all the Indian agents get rich. The first thing they do is to start a store; the next thing is to take in cattle men, and cattle men pay the agent one dollar a head. In this way they get rich very soon, so that

SHIPMENT OF 50 CARS OF CATTLE FROM WHITE PINE COUNTY NEVADA

Fig. 9. Long after Sarah Winnemucca's writing and activism, the expansion of the cattle industry continued. This shipment of fifty train cars of cattle through Nevada in 1927 would have confirmed her fears. Library of Congress, Prints & Photographs Division.

they can have their gold-headed canes, with their names engraved on them. The one I am now speaking of is only a sub-agent. He told me the head agent was living in Carson City, and he paid him fifteen hundred dollars a year for the use of the reservation. Yet, he has fine horses and cattle and sheep, and is very rich.[26]

By collecting rent from cattle ranchers on reservation lands that were not their own, these land agents make almost pure profit. Winnemucca emphasizes the mercenary focus of land agents, repeating the word "rich" throughout the passage. Interestingly, the most commonly used nouns in the passage are "cattle" and "agent." Linked through Winnemucca's parallel structure, where one action carries through to its complement, this equation overpowers the Paiutes, who can never see the riches of either agent or cattle. Indeed, Winnemucca declares that conflicts like the Bannock War are manipulated events designed to sell cattle and their products: "The only way the cattle-men and farmers get to make money is to start an Indian war, so that the troops may come and buy their beef, cattle, horses, and grain. The settlers get fat by it."[27] Both the reservation agents and the cattle ranchers take advantage of the U.S. policy that allows them to use cattle to quickly grow rich off the Paiute reservation lands, harming Paiutes themselves in numerous ways.

If Winnemucca demonstrates that cattle are a primary reason for the changes in Paiute territory, through removals and war, she also shows that the Paiutes cannot succeed even within the reservation system, as they attempt to assimilate to U.S. agricultural models. Cattle are catalysts for a form of subjugation that eliminates Paiute food sovereignty. In an interview published in the *San Francisco Morning Call* in January 1885, Winnemucca speaks of the Paiutes at Pyramid Lake. Cattle once again represent the loss of Paiute territory and the encroachment of U.S. settlers and their livestock. The only work the Paiutes have, Winnemucca tells the reporter, is occasionally when the men, like her brother Tom, "herd cattle in Summer or work on farms near the reservation."[28] The ranchers and corrupt agents collude once again: "They run their cattle on our reservation, driving ours and the game off. It is a wretched state of affairs."[29] Winnemucca also describes the meager provisions available to the Paiutes, who are forced to hunt as they receive too few government rations for survival. Instead of producing and preparing foods in their traditional manner, they are reduced to hunting whatever is available in a severely limited space. Rabbits must make up a large portion of all the game now available to them, and "you may imagine how quickly they will disappear when hunted by 7,000 starving Indians," she says.[30] The Paiutes are being denied recourse to their own systems of subsistence. Some version of this is similarly reported in January 1885 in the *Wheeling Register* and *Boston Daily Globe*, according to Carolyn Sorisio and Cari Carpenter, editors of the 2015 collection of Winnemucca's articles. The starving Paiutes have become manual laborers on some of the very farms and cattle ranches that have been created on land that should have been their own, that at one time had been their own, and that they could have been taught to turn into farms and ranches of their own. U.S. settlers had disrupted the lives of the Paiutes through their livestock almost as much as land agents manipulated the reservation system of land management. Unable to harvest, fish, hunt, plant, or prepare foods in their familiar lands, they must labor for wages or rations and make ends meet as they are able—until these efforts fail them, as well.[31]

Ultimately, the cattle boom of the 1880s meant that the Paiutes would lose out to the cattle ranchers and the reservation agents who capitalized on them. As part of the activism to return the Paiutes to some form of sovereignty, activism that included the publication of *Life among the Piutes*, Winnemucca visited Washington DC twice, in 1880 and 1884. In her first visit she traveled with her brother Natchez, her father, and two other men (see fig. 10). In this photograph from their visit, Winnemucca is dressed in U.S. Victorian "manifest manners" style, as are the rest of the family. She is not wearing the hybrid "Indian princess" costume of her lecture tour. Rather, she dresses in the "civilized Indian" style to impress the governmental officials she would meet with, to use Sorisio's definitions.[32] The Winnemucca family were not exactly official Paiute spokespeople, but U.S. government officials considered them useful representatives in this instance. In her

/   2   3   4   5

Fig. 10. In this photograph of the Winnemucca family's visit to Washington DC in 1880, Sarah Winnemucca (*left*) is accompanied by Chief Winnemucca, her brother Natchez, Captain Jim, and an unidentified boy. National Archives (75-1P-3-26).

April 1884 testimony to the House Subcommittee on Indian Affairs, Winnemucca describes actions and themes recognizable from her self-narrative. She directly asks for the Paiutes to be moved from the Yakama (spelled Yakima in her transcripts) Reservation to the McDermitt (spelled McDermot in her transcripts) Reservation, an area seemingly undesirable to cattle ranchers.

There, she argues, the Paiutes might be able to live unmolested: "Now my people say it is useless for me to ask for that Reservation back, and they say they would like to take anything you will give them and they will take this Camp McDermot, although Camp McDermot is worthless."[33] Winnemucca notes that by this point the Paiutes have been reduced to "nothing" and are at the stage of begging for a bare minimum subsistence. She continues: "We want that rather than to get nothing; rather than to be always and forever begging along the railroads. Anything is better than nothing."[34] With the help of her appeal, by 1889 the remaining Yakama-based Paiutes were moved to McDermitt, a reservation connected to the U.S. government as a military fort. Ironically, Winnemucca had a more positive relationship with the U.S. military than with the reservation land agents. Her strategies were most important when navigating the mismanaged and corrupt reservations (intolerable) and the more compassionately managed forts (preferable). While McDermitt was considered an undesirable location by the cattle ranchers and Paiutes alike, Winnemucca accomplished at least a minor victory for her people after years of determined work fighting against the cattle-based changes of her time and region.[35]

Winnemucca had to use the best tools available to her—her rhetorical skills—to challenge U.S. expansion and corruption, supporting her people as they moved from the Pyramid Lake Reservation, Nevada, to the Malheur Reservation, Oregon, and the Yakama Reservation, Washington Territory. These removals were all connected to expanding lands for U.S. cattle ranches, under a particular kind of settler colonialism in operation in U.S. policy toward indigenous communities during this time. As opposed to imperial conquest, people were moved onto reservations within the nation's borders and were supervised there. By the 1850s, when reservations started appearing formally, the United States no longer had the ability to remove individual tribes to a point beyond the nation's political boundaries; instead, the government created bordered spaces that worked to absorb tribes into U.S. society. People were coerced into remaining on these reservations—often by agents who actively manipulated their power, gaining profit at the expense of those located in bound spaces. Even lands that were not specifically obtained for cattle ranches were disciplined spaces for Native Americans. The United States maintained newly denoted wilderness spaces, such as Yellowstone National Park, with military force. Such reinforcement was another way to prevent the Paiutes and other communities from being able to sustain themselves through their traditional foodways. Rather, they were forced more strongly to learn Euro-American forms of agriculture. Government land allotments changed tribal cultures—including their cuisine and the production of foods—into a monoculture model connected to U.S. expansion and industrial agriculture. The impact of live-stock was largely felt in terms of the environmental changes brought about through agri-expansion. For the Paiutes, though, the impact of cattle was direct and devastating.[36]

In her writings, articles, and testimony, Winnemucca shows that U.S. ranchers manipulated land policy to acquire greater territory at the expense of Paiute lands and their traditional food-ways. Indigenous foods in stories have become a necessary com-ponent of environmental justice: the complicated relationship between Native American culture, its representation, and ecology and ecological discourse. Native American nations and tribes in many ways are under threat of cultural misappropriation as much in the twenty-first century as during Winnemucca's time, as she manipulated the stereotype of the "Indian princess" to her own designs. The issue remains that the Paiutes were driven out of their home territory for cattle ranches. In so doing the U.S. settlers not only decimated the Paiutes and their culture but also caused great ecological damage to what is now the U.S.

West, damage that has continued with nuclear testing and other environmental issues in the region. Tradition, preservation, and sovereignty risk being uprooted, if not for the literature and the ongoing activism that continues today, a legacy that Winnemucca helped develop through her rhetorical strategies on food, justice, and cattle. She resisted the situation presented to her people through nearly her entire adult life, and one of her primary concerns throughout was the graphic representation of the harsh treatment of Paiutes by U.S. settlers and land agents, treatment that was dehumanizing in nature.[37]

Treated Like Cattle

Some of Winnemucca's most alarming rhetorical uses of cattle occur when she associates them with the Paiutes themselves. During the cattle boom era, the Paiutes were being completely dehumanized by ranchers and agents. In *Life among the Piutes*, for example, Winnemucca compares the commodified cattle and other livestock with the now-commodified, wage- and ration-dependent Paiutes: "We were turned over to him [James H. "Father" Wilbur] as if we were so many horses or cattle. After he received us he had some of his civilized Indians come with their wagons to take us up to Fort Simcoe. They did not come because they loved us, or because they were Christians. No; they were just like all civilized people; they came to take us up there because they were to be paid for it."[38] Here, the U.S. manner of dealing with land and other resources is based entirely on capital—usually cash, but also commodities that could be turned into ready cash, like cattle. The Paiutes are treated exactly like "so many horses or cattle." These animals have value as commodities only—or what they could produce or the work they could perform—and the Paiutes are treated the same. Moreover, the "civilized Indians" are paid currency for their labor—and economics are the only way in which things—or people—have value in

Euro-American agri-expansion. Winnemucca ironically compares this commodification to the community's notion of civilization, a move that removes all Paiute sovereignty. Winnemucca uses cattle as a metaphor throughout to frame questions of biopolitcs in the Foucauldian and Derridean senses. The Paiutes are shown being driven out of their home territory so that cattle ranchers—and their stock—can move in, and in so doing those settlers harm humans and nonhumans alike. Winnemucca gains sympathy with her audiences by showing striking parallels between the treatment of Paiutes and cattle.[39]

In her 1884 testimony, Winnemucca reiterates her argument from *Life among the Piutes*, in terms of her people being treated as cattle, as less than human. She describes the transition from the tolerable agent, Samuel Parrish, to the corrupt and cruel one, William Rinehart, comparing Paiutes to cattle in terms of their treatment by U.S. settlers once again: "Our good man turned us all over to this man, so many little children, so many women, so many old men, so many young men—we were turned over to him just the same as you would turn your stock over to a person who bought them."[40] In addition, there is a description of the Paiutes being forced to move to Yakama, to another corrupt agent, James H. "Father" Wilbur, mentioned above. Winnemucca describes this fatal move, a three-hundred-mile trip in winter: "My peoples dead bodies were strung all along the road across the Columbia River to this Yakima Reservation. When we got there we were turned over to another man, and then after we got there we died off like a lot of beasts, and of course then the following winter I came right here to Washington."[41] One final time, and similar to her discussion in *Life among the Piutes*, Winnemucca turns to her symbol of cattle (in this case "beast" or "stock") to describe the treatment of the Paiutes by the agents. The Paiutes are no longer forced off the land to make room for cattle, nor are they starving over the lack of proper rations; now they are dehumanized entirely. The only move left to her, Winnemucca declares, is to travel to Washington to make a strong appeal for aid.

In a *Baltimore American* article of January 1884, during her second Washington visit, Winnemucca describes events leading up to one of her people's removals. The theme of cattle is highlighted again, but in this case Winnemucca describes the Paiutes' failed attempt to be cattle ranchers, caused by the changes in reservation and leadership structures: "The new man came July 1st, and that fall all my people were sent away from their reservation, and they were starving. The five hundred head of cattle put on our reservation for us were scattered and gone, and my people got none."[42] Paiutes are compared to cattle in this example, as both two-legged and four-legged creatures are "scattered and gone" or "sent away" from the reservation. Once again, Paiutes have not been allowed to succeed in the corrupt system, but now Winnemucca shows a different example of the wrongs against her people than in previous writings and testimony. In this account, she displays the efforts of the Paiutes to emulate the Euro-Americans through working within the structure of cattle ranches, emphasizing how this fails as well through her parallel narrative structure.

Winnemucca does not stop there, either; she continues to give newspaper interviews after her self-narrative and testimony, always hoping to help her people—again using cattle as an extended metaphor. In the *Morning Call* article discussed earlier, Winnemucca says: "The Piutes are on the verge of starvation. They are growing weaker and weaker every day for want of food. They have been driven like wild beasts from place to place, and forced back from the meadows and the banks of rivers and streams into the mountains that are barren and wholly destitute of game."[43] While being marginally better for them than Yakama, life for the Paiutes at Pyramid Lake remains unforgiving. They are weakened, starving, and under the control of corrupt land agents as usual—this time one named William Gibson. Indeed, Winnemucca's brother Tom struggles and dies at Pyramid Lake instead of moving to Yakama. He is doomed either way—his ending almost as unchangeable as that of the domestic cattle destined for slaughter.

The significance of using cattle as representing food sovereignty and the Paiutes themselves appears as well in the work of Winnemucca's brother Natchez, quoted in *Life among the Piutes*.[44] He uses cattle both as symbol of devastating change as well as comparably tormented humans like the Paiutes. Addressing Secretary of the Interior Carl Schurz, Natchez writes: "O, good Father, it is not your gold, nor your silver, horses, cattle, lands, mountains we ask for. We beg of you to give us back our people, who are dying off like so many cattle, or beasts, at the Yakima Reservation."[45] Natchez seeks to communicate the treatment of the Paiutes as an epidemic that might run through a cattle herd. Perhaps this "civilized" audience can show concern about the plight of the Paiutes if he frames his argument in the language of commodified livestock and investments. Winnemucca and Natchez find it impossible to communicate clearly on the basis of shared humanity, so they turn to metaphoric language using cattle and capital. As cattle can be protected from dying of disease, so, they argue, can the Paiutes. Natchez frames his narrative grammatically to "silver, horses, cattle, lands, mountains." In each case, the commodity is either currency or an item that can be easily transformed into currency through some kind of extractive economic structure. Interestingly, everything on Natchez's list is part of the natural world—linking the exploitation of the Paiutes to that of nature through U.S. expansion, as well.

Fittingly, *Life among the Piutes* takes its place among preeminent works by Native Americans; Winnemucca's self-narrative is the first "ethnoautobiography by an Indian woman," according to Ruoff, one that "has more in common with the life histories written by women of color and of the working class and with the post–Civil War autobiographies of African Americans than with those by mainstream middle-class women."[46] In addition, LeAnn Howe describes this writing as "tribalography": "Native people created narratives that were histories and stories with the power to transform."[47] Leah Sneider defines tribalography as "part autobiography, part memoir, part history, and perhaps even part fiction."[48] *Life among the Piutes* challenges standard genre forms. A self-taught writer, Winnemucca never perfected the nuances of writing in English. To a large extent she relied on her editor, Mary Peabody Mann, for issues of grammar and spelling. Mann insisted, though, that the ideas and presentation of the work were Winnemucca's alone. Winnemucca maintained her ability to play on metaphor and rhetoric, investing key symbols throughout her work, specifically cattle, with great significance. While Elizabeth Peabody was "the premiere purveyor of language theory" among the Transcendentalists in terms of "metaphorical language," according to Philip F. Gura, the Winnemucca siblings' shared experiences influenced their work as well.[49] I argue that her rhetoric also has a posthuman element. Not only do Natchez and Winnemucca compare Paiutes to Euro-Americans; they also compare humans to cattle. While the Peabody sisters worked with Winnemucca to craft some of her written rhetoric, much of her style was developed through performing and working with her family over several years—in this case, a style deeply engaged with the metaphor of cattle and the need for greater Paiute sovereignty.[50]

Thus, Winnemucca joins a lineage that began with Pocahontas and Sacajawea as noteworthy Native American women in U.S. culture. While Zanjani argues that Winnemucca is not as famous as Pocahontas and Sacajawea because she did not directly assist Euro-American males, Winnemucca's truest legacy is in her rhetoric.[51] Like her predecessors, Winnemucca came of age at a time of dramatic change in the United States, as she was forced to navigate the loss of territory and the incursion of U.S. settlers into her homeland. She worked within a difficult and shifting historical context in U.S.–Native American relations, practicing the concept of "survivance (survival + resistance)" in her dealings with government representatives, according to Powell.[52] Winnemucca navigated an impossible situation with as much grace as possible, trying to bridge two cultures and worlds. In this way, her writing has a survivance presence—it lasts and continues to be deeply felt.[53]

Winnemucca's writing compares cattle and U.S. settler colonialism in some of her strongest arguments for Paiute food justice, and in so doing it leaves a legacy of struggle for sovereignty that resonates—perhaps in some ways more strongly—to this day. Winnemucca draws striking parallels between Paiutes and cattle, as well as showing how cattle helped destroy Paiute foodways. In the late twentieth and early twenty-first centuries, food sovereignty has returned to international policy debates, as seen in the 1996 World Food Summit; it has been made more direct during the First Indigenous Peoples' Global Consultation on the Right to Food and Food Sovereignty organized in 2002 by the International Indian Treaty Council, made up of indigenous peoples across the Pacific, the Americas, and the Caribbean. Winnemucca accomplishes this level of association through her rhetorical strategies and deliberate use of cattle as catalyst for issues of Paiute food justice. She seeks to bring an element of traditional land policies—specifically those involving food justice—back to her people at a time of U.S. expansion. Her attention to stylistic choices in her narrative makes her argument for Paiute land and food sovereignty against the U.S. cattle ranches one of the most enduring narratives of the nineteenth century, and according to Zanjani, "moved her father, Chief Winnemucca, to declare to his people at the end of the Bannock War, 'None of us are worthy of being chief but her.'"[54]

Conclusion: The Ongoing Struggle for Food Sovereignty

Winnemucca's critiques of U.S. land policies against the Paiutes and her use of the metaphor of cattle to show the treatment of Paiutes place her alongside other activists of her time. However, Winnemucca has always been a controversial figure. She gambled, drank, and had multiple marriages and relationships.[55] Many Paiutes also perceived her as being too assimilationist. In a further complication, the Paiute Wovoka, or Jack Wilson, began the Ghost Dance ritual during the time that Winnemucca was writing and lecturing. The charismatic Wovoka experienced revelations, according to biographer Michael Hittman, including the return of the dead and "the practice of a round dance, which would also effect an earth cataclysm and so result in the removal of white men."[56] Wovoka promised revenge against the U.S. settler-colonists, a return to familiar agricultural methods, and reunion with lost Paiute loved ones. He made a powerful figure at the time of U.S. industrial agriculture, as the nation's capitalism resulted in tragedy for the Paiutes and other nations and tribes. Like Winnemucca, Wovoka had early experience with industrial agriculture and market-based commodities as a form of oppression by the U.S. government beginning as early as 1870. Both were forced to negotiate an untenable system of agriculture, and Wovoka likewise reacted powerfully against this oppression. While Wovoka developed the Ghost Dance as a desperate action against cultural annihilation, Winnemucca had also been engaged with bettering the lives of her people throughout her writing and speaking career. Michael Omi and Howard Winant discuss Wovoka's Ghost Dance, along with Geronimo and Sitting Bull, as examples of "resistance culture."[57] To these I would add Winnemucca, who worked through her rhetoric to ameliorate—rather than wholly reject—the reservation system. Perhaps this modification had to do with her family's history of working with U.S. settlers, as seen with Truckee and Chief Winnemucca. Perhaps, as well, she worked within her known limits—realizing what was possible or likely to be achieved during her lifetime. In this, she was optimistic. The Dawes Allotment Act of 1887 created further land woes for the Paiutes and was seen as a final betrayal. The reservation system was in place by her time and only grew worse during her career; thus, she may have realized that making some material change within the established system—moving the Paiutes to McDermitt and away from Yakama, for example—was one of the best goals she could accomplish for her people.[58]

After Winnemucca's death in 1891, at age forty-seven, a number of sympathetic obituaries were written regarding her work on behalf of the Paiutes, noting the connection between cattle, U.S. expansion, and Paiute suffering. The *New York Daily Herald* wrote: "Her main purpose in life was to have the Piutes removed to a reservation far from the cattle settlements of the whites, where they couldn't get liquor, and to have the Government give them aid in farming, but she did not live to accomplish this work."[59] Even a relatively dispassionate entry like this obituary in the *Daily Herald* makes the point that the encroachment of U.S. settlers and their cattle into the western lands of the continent during the U.S. cattle fever was disastrous to all that lived there, in this case the Paiutes. Winnemucca spent her life using her skills in rhetoric to demonstrate to people across the United States that cattle and the U.S. settlers who brought them were metaphorically and literally destroying the Paiutes. In so doing, she became, according to Ruoff, "the Paiutes' mightiest word warrior."[60] Interestingly, at almost the same time that Winnemucca was writing and performing, María Amparo Ruiz de Burton was writing one of the earliest full-length works by

Mexican American women. She also engages with the symbol of cattle in her work on colonialism and U.S. agri-expansion in the West during the years following the Treaty of Guadalupe Hidalgo, complementing Winnemucca's work. In the next chapter I analyze Ruiz de Burton's novel *The Squatter and the Don* (1885), which engages with cattle, land rights, and the people struggling over them.

FIVE

The Cowboys *Are* Indians in
The Squatter and the Don

"I don't want any cattle. I ain't no '*vaquero*' to go '*busquering*' around and *lassooing* cattle. I'll *lasso* myself; what do I know about whirling a *lariat*?" said Mathews.

"You will not have to be a vaquero. I don't go '*busquering*' around *lassooing*, unless I wish to do so," said the Don. "You can hire an Indian boy to do that part. They know how to handle *la reata* and *echar el lazo* to perfection."

—RUIZ DE BURTON, *The Squatter and the Don*

In one of the most analyzed scenes from María Amparo Ruiz de Burton's *The Squatter and the Don* (1885), the protagonist, Don Mariano Alamar, admits that the Native Americans workers on his rancho are capable of "perfection" when it comes to the subtleties in the craft of cattle ranching, the iconic art of the vaquero (in English, cowboy). They ably work the "reata," lariat, and "lazo," lasso, for example. Ruiz de Burton emphasizes with the use of italics several words specifically associated with being a cowboy. *Vaquero*, for example, is almost a direct Spanish-to-English translation for "cowboy" or "cattle driver." While "lasso" is known to this day, the "lariat" is the rope itself, and "busquering" is an

Anglicization of the verb *buscar*, to look for. Although the rancheros are the elite Californios/as—descendants of European colonizers in the former Mexico—the indigenous peoples of Mexico and the United States are at least as skillful with the items largely associated with the cowboy tradition of U.S. cultural romance. Indeed, cattle and the people who work with them are important symbolically and materially over the course of the novel.

Cattle are the driving factor in the most dramatic scenes of *The Squatter and the Don*, and the Native American characters are seen working with them in each of these scenes; therefore, they are also significant figures even beyond what has perhaps been discussed so far in Ruiz de Burton criticism. The indigenous peoples of the San Diego County/northern Baja California region are the Kumeyaay (also called Diegueño).[1] In the novel, the Kumeyaay perform much of the technical and skillful work with cattle, the true cowboys or vaqueros. The primary conflict of the novel sets the "Don" against the U.S. settlers, named "squatters" in the text, over land rights. Indeed, Ruiz de Burton's focus on the issue of "squatters" is reminiscent of Washington Irving's and James Fenimore Cooper's use of the term that we saw in chapters 1 and 2. The lands that the two nominal landowners are fighting over are cattle ranchlands, and the vaquero terminology is as integral a part of the novel as Ruiz de Burton's use of legal documents.[2] Looking at *The Squatter and the Don* specifically, therefore, I argue that through their skill with domestic cattle—and the obvious importance she affords cattle as part of the economy and culture of California pre-and-post 1848 and the Treaty of Guadalupe Hidalgo that ceded California to the United States from Mexico—Ruiz de Burton cannot help but show value to the vaqueros, though they largely inhabit the periphery of her novel. I maintain that she uses these seemingly marginal Kumeyaay characters in distinct ways, that her depictions of these characters are racially charged and show her racism as a Mexican

American woman of European descent, and yet, despite this, she perhaps unintentionally depicts them as the most capable and skilled workers with cattle. As the use of cattle and agriculture in the novel is Ruiz de Burton's focus for her argument about the need for land rights and sovereignty, and as she neglects to engage with indigenous land rights, the centrality of these characters complicates the more direct contrast she makes between Californios/as and U.S. settlers.[3]

Ruiz de Burton's manner of playing with contrasts, expected stereotypes, and parallel figures through the Californio/a and squatter characters opens up such readings to her Kumeyaay characters as well. The primary story line in *The Squatter and the Don*, again, has to do with land rights and cattle. The novel tells the story of two families: the Californio/a Alamars and the U.S. American Darrells. The Alamars lose everything over the course of the novel due to the manipulation of land laws that in effect allow U.S. squatters like the Darrells to legally appropriate lands from families like the Alamars. These antagonists become connected through a love story, as the Alamar daughter, Mercedes, ultimately marries the millionaire son of the Darrell clan, Clarence. Don Mariano Alamar is a Californio elite who has had to reassess his citizenship and privilege within a new nation after losing his land to the U.S. squatters. He had originally been deeded a 47,000-acre land grant that was under siege by those squatters who manipulated a legal loophole, the "no-fence" law of 1851. This law stated that new settlers were not required to fence their newly purchased lands to keep the cattle from their crops, which allowed farmers to sue for damages when their crops were then eaten by free-ranging cattle. This was one example of the fundamental difference in agricultural styles of the Spanish (ranching, relatively passive) and the English (intensive, agrarian), a distinction that would remain into the nineteenth century in the expanding United States. The squatters of the title refers to a role the Treaty of Guadalupe Hidalgo made available for U.S. settlers interested in gaining land in the West, whereby they could "squat" on Californio/a lands, residing there extra-legally. As long as they worked the land, they could apply for the legal rights to own that land, the original titleholders of the contested rancheros then having to sue to maintain their legal ownership. Such a process was expensive and time consuming, often taking years, and most Californios/as, like Don Mariano in the novel, lost their lands and stock in the process. Ruiz de Burton declares: "This was, of course, ruinous to Don Mariano. . . . Now any one man, by planting *one acre* of grain to attract cattle to it, could make useless thousands of acres around it of excellent grazing, because it became necessary to drive cattle away from the vicinity of these unfenced fields."[4] Therefore, even though these new citizens to the United States had been granted the right to maintain their lands following the Mexican-American War, they often lost it anyway.[5]

Cattle became a significant device in this legal battle, as the squatters were allowed to kill any that trespassed onto their unfenced wheat fields. In *The Squatter and the Don*, Alamar's cattle are killed in land disputes, and eventually those that remain are sold and moved to pay Alamar debts in one of the novel's tragic scenes. Ruiz de Burton's narrator later states sarcastically: "Thus, every night the fusilade of the law-abiding settlers would be heard, as they, to protect *their 'rights under the law,'* would be shooting the Don's cattle all over the rancho."[6] This scene echoes another one of wasteful killing by U.S. settlers on the frontier, the decimation of the bison of the Great Plains region. In both cases, there is no purpose to the killing of these different kinds of cattle other than to establish a certain kind of ownership of terrain. The meat, milk, or other goods, even the value of the animals themselves, is of no concern. The wastefulness of the U.S. settlers to the region is remarkable in both instances. In several key scenes in the novel, Ruiz de Burton focuses on this wasteful killing of cattle.

Thus, the Alamar rancho is the site for the main action of the novel, and the workings of a cattle ranch are omnipresent throughout the legal plot against which the love story occurs. Indeed, *The Squatter and the Don* is one of the U.S. nineteenth-century novels to deal most directly with vaqueros and the workings of a cattle ranch, having the most dramatic instances of the cowboy life of any of the books discussed in *Cattle Country*. In almost every case, interestingly, the cowboys are Kumeyaay. While the Kumeyaay characters are not involved in the romance story line, their presence is felt in the novel's scenes most directly connected to working with cattle. The Kumeyaay of *The Squatter and the Don* do not have as much voice as either Ruiz de Burton's Californio/a heroes or Sarah Winnemucca Hopkins's people, the Paiutes. Where the Paiutes are punished for "hunting" horses and cattle within the fenced terrain of the U.S. Great Basin of the 1870s, Ruiz de Burton's Californio/a characters suffer because the U.S. settlers refuse to fence in their wheat fields. These settlers find a way to legally kill the ranchers' cattle—because those ranchers have been made into something less than citizens following the Treaty of Guadalupe Hidalgo. Don Mariano seeks to preserve his cattle from the U.S. squatters, and the Kumeyaay are the people who most directly help him care for them. The Kumeyaay work while other characters worry about their status and sovereignty; indeed, to a large extent the Kumeyaay work *so that* other characters have the ability to worry about their status. I apply the structure of Marcial González's analysis of *The Squatter and the Don* to frame my discussion of the varying degrees of labor and privilege used by Ruiz de Burton in the scenes of the vaquero agricultural model. I also follow José Saldívar's reading of this as being a sentimental historical novel "with a Californio difference."[7] Ruiz de Burton's version of sentimentalism allows the Kumeyaay to have an interesting role in the novel, significant in their survivance against a background of double colonization.[8] González names the Kumeyaay labor of *The Squatter and the Don* largely invisible, but I argue that they are in fact regularly seen working with the cattle—the novel's most important symbol. Whether or not Ruiz de Burton means for us to focus on the Kumeyaay characters in the novel, they are shown performing some of the story's most interesting vaquero labor: working with the land, expertly using the lassos, and driving cattle.[9]

The Kumeyaay, the indigenous tribe of that region, are described in the novel in a somewhat liminal sense; indeed, they are rarely mentioned at all, and almost always disparagingly. Ruiz de Burton's racial bias, at work in this novel, has been well discussed.[10] In addition, the Californios/as and Kumeyaay can be seen as representing work as opposed to labor, where work involves ownership of material items, as opposed to labor, which does not translate into ownership of any capital. The Californios/as own the land and the cattle until they are defeated by U.S. squatters, while the Kumeyaay work with and for them on that land; theirs has often been considered a lesser labor to the Alamar family's work. As much as Ruiz de Burton endeavors to gain sympathy for her own people, the Californios/as, she fails to spend an equal amount of time on the Kumeyaay characters. At the end of the novel, both groups are proven to be victims of the U.S. settlers and railroad monopolists. The marginalized peoples become partnered—if to different degrees—in their treatment by the U.S. Americans. Although Ruiz de Burton uses demeaning language to describe her Kumeyaay characters, her demonstrations of their actions show their skill in one of the most romanticized professions in U.S. culture. In addition, Priscilla Ybarra writes of Ruiz de Burton's work in terms of her environmental concern and issues of race and ethnicity as a kind of "goodlife writing" as opposed to a limited view of environmental writing.

This negotiation of cultures, environment, and cattle show Ruiz de Burton's complicated navigation of what it means to be native in California in the nineteenth century. Indeed, Don Mariano (and by extension Ruiz de Burton herself) seems to admire the Kumeyaay increasingly as the novel progresses. As the Don loses his own privilege at the hands of squatters like William Darrell and railroad monopolists like Leland Stanford, he comes to consider the Kumeyaay more favorably by contrast.[11]

The vaquero history of California predates and makes an important historical context for *The Squatter and the Don*, as Ruiz de Burton both criticizes U.S. policy and highlights California cattle culture of the second half of the nineteenth century through these integral characters. Tereza Szeghi notes Ruiz de Burton's use of the Spanish word *vaquero* rather than the English *cowboy* throughout the text, though the English word would have been recognizable by the 1880s.[12] Ruiz de Burton emphasizes the importance of Mexican American or Californio/a culture in the nineteenth century through her vocabulary. The Mexican-American War was one of the vaquero culture's entry points into U.S. history. Spanish and Mexican people influenced the U.S. West, perhaps most significantly, by teaching the vaquero skills to what would become the cowboy. The vaqueros of California were making a U.S. version of the Latin American subculture of cattle ranching. California's specific vaquero history began in 1769 with the Franciscan missions of Father Junipero Serra's priests, along with their Native indentured servants. Among other charges given these unpaid workers was the responsibility for tending the livestock, labors they came to perform with great skill, becoming expert with horses and cattle (see fig. 11). In the century leading up to the time described in *The Squatter and the Don*, indigenous nations and tribes throughout California—like the Kumeyaay—were learning to be skilled vaqueros, working with lassos and other tools, even though they learned these skills through oppression and slavery.

In addition, the Indenture Act of 1850 and the Greaser Act of 1855 both were used as a form of slavery—of forcing Native Americans in California into unpaid labor for at least a decade. This history informs the Kumeyaay characters in the novel. Although they are not given high status in *The Squatter and the Don* for their vaquero labor, Ruiz de Burton provides a place for readers to extend their status to one of prominence through her coding of cattle, the value cattle have for the Californios/as, and, by extension, the value of those who can sustain and preserve them.[13]

The Mexican vaquero history also appears through the cattle themselves. Don Mariano has Mexican stock cattle, presumably. Meanwhile, one of the squatters, Miller, notes that he is concerned about his English "milch cows" that were brought west upon settlement. One of the U.S. squatters uses his concerns about getting the milk cows in by sunset as a motivator for an important meeting in the novel. At this meeting, Don Mariano replies, "Exactly, we want to look after our cows, too."[14] As this is a casual scene establishing a comfortable opening for the meeting, the language here is lighthearted and polite—especially as concerning the Don: "All saw the fine irony of the rejoinder, and laughed heartily. Miller scratched his ear, as if he had felt the retort there, knowing well, that with the exception of Mathews and Gasbang, he had killed and '*corraled*' more of the Don's cattle than any other settler."[15] Don Mariano's joke shows his excellent English and good manners, but it also pointedly demonstrates why he is there in the first place. His grace is contrasted to the squatters' lack thereof. The character Mathews compares the Californios/as to cattle, or other livestock, in an earlier scene: "Those greasers ain't half crushed yet. We have to tame them like they do their mustangs, or shoot them, as we shoot their cattle."[16] Ruiz de Burton shows Mathews to be a reprehensible character, speaking of humans as if they were animals. Such a move by Ruiz de Burton allows the readers to sympathize with

Fig. 11. *"Toro, toro!,"* In this classic 1893 image, Frederic Remington shows what would become one of the most iconic figures of the West, the cowboy. Library of Congress Prints & Photographs Division.

her Californio/a characters, like the Alamar family and, by extension, the Kumeyaay, both sets of characters being poorly treated.

This leads to one of the most analyzed passages in the novel, the crux of Don Mariano's plan to save his land from the squatters and capitalize on a new opportunity. "All I want to do," he says, "is to save the few cattle I have left. I am willing to quit-claim to you the land you have taken, and give you cattle to begin the stock business, and all I ask you in return is to put a fence around whatever land you wish to cultivate, so that my cattle cannot go in there."[17] The quit-claim idea that Don Mariano describes involves leasing his land to the squatters, asking only that they fence their lands in return, thus removing the squatters' excuse for killing his cattle. In addition, fencing land would have been made easier and cheaper through Joseph Glidden's patent of barbed wire in 1874.[18] Unsurprisingly, the U.S. settlers do not listen Don Mariano's suggestions, though they are well reasoned and come from his lived experience in the region. Clarence Darrell, a key hero figure of the novel, interprets the "no-fence law" that the U.S. squatters manipulate so effectively: "It is worse than that [iniquitous], it is stupid. Now it kills the cattle, afterwards it will kill the county."[19] In fact, instead of raising cattle, they choose to grow wheat on a large scale, though this is a less tenable agricultural system and will eventually lead to railroad monopolies and the tragic events detailed in novels like Frank Norris's *The Octopus* (1901).[20]

Ruiz de Burton's own history demonstrates why readers should pay more attention to the Kumeyaay vaqueros who perform the labor that has been sentimentalized throughout U.S. literature and culture (see fig. 12). Her husband, Captain Henry S. Burton, was stationed in San Diego County and was specifically assigned to "control native peoples."[21] These actions resulted in the decimation of indigenous California populations. California-based Native Americans numbered between 70,000 and 150,000 in 1848, and their numbers dropped to 30,000 by 1869 and to less

Fig. 12. María Amparo Ruiz de Burton, seen here in an undated image, fought throughout her adult life to maintain her Rancho Jamul against the incoming U.S. squatters, much like her character Don Mariano Alamar. Graves Family Collection.

than 20,000 in 1900.²² The Luiseño and Kumeyaay tribes would have been closely connected regionally in the southern California area where Burton was stationed and where Ruiz de Burton would acquire Rancho Jamul. She was herself embroiled in legal action for Rancho Jamul for much of her adult life. In a sense, her real-life battle is dramatized in the novel, as is the similar struggle of her friend Mariano Guadalupe Vallejo, recognized as a model for Don Mariano Alamar.²³

At the same time, Don Mariano gives a speech showing his contempt for his Kumeyaay workers:

The land-owners were useful in many ways, though to a limited extent they attracted population by employing white labor. They also employed Indians, who thus began to be less wild. Then in times of Indian outbreaks, the land-owners with their servants would turn out as in feudal times in Europe, to assist in the defense of the missions and the sparsely settled country threatened by the savages. Thus, you see, that it was not a foolish extravagance, but a judicious policy which induced the viceroys and Spanish governors to begin the system of giving large land grants.²⁴

In one of his most feudal statements, the kind of commentary that has made contemporary critics read Ruiz de Burton as employing the sentimental genre popular during her time, she allows Don Mariano to speak positively of the Spanish and Mexican mission system that spread from Florida to California. The Don appreciates the "civilizing" element of the missions and their religious and cultural influence on the "savage" Native American populations, making them "less wild." Furthermore, he euphemizes "employed" instead of the slave system of labor that was actually in place throughout the mission structure. This highlights the complicated work Ruiz de Burton does with her Kumeyaay characters—denigrating them through a character like Don Mariano while repeatedly showing their proficiency as vaqueros, one of the most romanticized skill sets of the U.S. West.

Much of Ruiz de Burton's other work supports claims of her inherent racial stratification in her life and writings. For example, she was friends with the wife of Jefferson Davis, and they are known to have mocked U.S. northerners together. Don Mariano makes his impassioned economic proposal to the squatters because history has truly changed for him. Vincent Pérez observes the statistics of Mexican, as opposed to U.S., settlement of California in the nineteenth century. In 1841 between 6,500 and 7,000 Mexicans—and fewer than 400 inhabitants from all other nationalities—resided in what would become the state of California. In 1850 there were 10,000 Mexicans (of which only 3 percent owned large ranches); at about that time, Mexicans made up only 11 percent of the state's population. So, although the Mexican population had grown following the Treaty of Guadalupe Hidalgo, the U.S. population had exploded by an order of magnitude. That shift continued through the second half of the nineteenth century. Between 1860 and 1900 the population of the state rose from 380,000 to close to 1.5 million, while Mexicans made up no more than 2 percent of that number.[25] The sheer bulk of the incoming population gave U.S. settlers, or squatters, a privileged position in terms of legal battles and land-use policies. Jesse Alemán observes that at one point, Victoriano Alamar wishes that his family were squatters, as they seem to have more rights, to get away with more than the Californios/as do: "Ironically, Victoriano, and *The Squatter and the Don* as a whole, seems to 'forget' that the Alamars, and Californios in general, are themselves 'squatters' on Indian-owned lands."[26] Indeed, the Alamar family doubly profits from settler colonization: first through Spanish colonization before the events of the novel, and second through the merger through marriage of the Alamar and Darrell families.

Truly, three marginalized groups struggled for a place within U.S. society in the second half of the nineteenth century in California: newly emancipated African Americans, Mexican Americans, and Native Americans. Ruiz de Burton focuses her energies nearly exclusively on Mexican Americans, spending far less time on other characters. Meanwhile, I focus in this chapter on the Native American characters in her work. Throughout the novel, therefore, I analyze the ways in which Ruiz de Burton contrasts the Kumeyaay—her true vaqueros—to the Californios/as, the U.S. squatters, and finally the railroad monopolists represented by Leland Stanford. In these cases, although Ruiz de Burton speaks little about the actual Kumeyaay doing the vast majority of work in the novel (and receiving practically no credit for it), they are there to be seen when we readers look for them. They are the very cowboys that would appeal to audiences from Ruiz de Burton's time to the present. While she does not become another Herman Melville, writing an ode to labor and laborers as in *Moby-Dick* (1851), elements of Kumeyaay skill and value, and the romance of the cowboy, appear throughout *The Squatter and the Don*. Saldívar enthuses that the novel is replete with enticing drama: "*vaquero* cattle drives across mountains and rivers of blood, violent confrontations over cattle-raising versus farming etiquette."[27] The Kumeyaay maintain the cattle, are expert at rope work, and survive and thrive in situations that destroy the Mexican Americans and other characters. They are the closest Ruiz de Burton comes in her novel to the image of a true cowboy, of particular significance in a work focused on the appropriate usage of the land and environment as well as the keeping of cattle.

The Squatter and the Vaqueros

In *The Squatter and the Don*, wealth is connected to the environment. The proper use of land, and those who perform the labor on that land, is a primary theme of the novel. Unsurprisingly,

Don Mariano's use specifically of a hearty breed of Mexican cattle on his ranchero is described as the appropriate agricultural use of resources for both the land and for the humans working the land. While ecocritics tend to write about more recent literature, Ruiz de Burton's nineteenth-century novel engages with many of the same issues of resource management, waste, agriculture, and land development. However, Ruiz de Burton is not writing to challenge a system of power, entirely. She is only concerned about the change of power—from the Californios/as to the U.S. settlers. Indeed, the novel is focused on an economic issue throughout, rather than a more traditional romance. Cattle were an important part of California history, changing California's environment as part of European settlement and imperialism on large cattle ranches. The cattle were both consumers and consumed at the fulcrum of cultural politics in nineteenth-century California. It is curious that with significant attention being spent on Native Americans in U.S. literature, more attention has not been spent on them within this novel, especially in their role as cowboys; indeed, it seems as though readers are invited to do this work. Against the legal battle of the novel, the cattle rancheros worked by the Californios/as and Kumeyaay characters are where the majority of action occurs.[28]

In one of the most dramatic scenes in the novel, Ruiz de Burton takes her intention to its logical conclusion. The patriarchs of the rival families come to physical violence in a chapter that also hints at the culminating battle with its title, "The Squatter and the Don." One primary character insults the other too deeply, and violence follows. Dehumanizing language in fact becomes dehumanizing treatment. Like the other primary plot of the novel, this violence occurs between the two Euro-American families. The actions and reactions of the Kumeyaay characters are relegated to the periphery. The scene ends when patriarch William Darrell is actually lassoed as if he were an ill-behaved bull. It is represented as a fitting punishment, since he had just called Don Mariano's daughter Mercedes a "filly" being "paraded" to attract eligible young men for marriage. Adding to this insult, Darrell attempts to hit Don Mariano with his whip—treating the Californio like a beast. This is the impetus for the Alamar sons to act in protection of their father. First, Don Mariano himself displays his skill with horse riding. "The horse jumped aside, sat on his haunches for an instant, half-crouching, half-rearing, and in a second he was up again." Meanwhile, Darrell has only "clumsy horsemanship."[29] Enraged, he continues to attack, until he is thwarted:

> Darrell followed close, and again lifted his whip to strike, but instantaneously he felt as if he had been struck by lightning, or as if an aerolite had fallen upon him. His arm fell powerless by his side, and an iron hoop seemed to encircle him. He looked down to his breast surprised, and there the coil of a *reata* held him in an iron grip, and he could not move. He looked about him amazed, and saw that the other end of the *reata* was neatly wound around the pommel of Gabriel's saddle, and that young gentleman sat quietly on his horse, as if waiting Mr. Darrell's orders to move, his handsome face a little pale, but quite composed.[30]

Gabriel Alamar is defending his father in this scene—and displaying his lassoing skills as well. More importantly, though, this experience enrages the Darrell patriarch to the breaking point. He never fully recovers. He calls for his "pistols" to be brought so that he can hurt the "greasers." His own bigotry is now on full display in his embarrassment at being out-skilled at horsemanship and cattle driving by the Alamar men.

During the scene, "two Indian *vaqueros*" appear and are entertained by the spectacle. They witness the violence on display as something potentially done in sport: "Noticing that the patron,

Don Gabriel, held a *reata* in his hands, the *lazo* end of which was attached to Darrell, they thought that for sport Don Gabriel had thrown the *lazo* on the old squatter. Having come to this conclusion, they began to shout and hurrah with renewed vigor."[31] In a sense, this scene appears to be a version of the displays of skill that would become famous in U.S. culture in rodeos and other cowboy shows. However, in this scene the vaqueros are supposedly viewed as misunderstanding the truth of what they witness. They shout encouragements in Spanish and are shamed for their lack of knowledge of English—though there is no way for them to have learned English by this point. In Spanish, therefore, they half-mockingly correct Gabriel's technique: "Apriétate viejo! apriétate míralo! ya se ladea!" (Tighten up, old man! Look at that! It's already tilting!).[32] Again, Ruiz de Burton historically describes her Native American characters in ways that have been analyzed much in criticism.[33] Critics have noted the contrast between Don Mariano's perfect English—much is made of it in an earlier scene—and the vaqueros' lack of English knowledge. In fairness, though, the Kumeyaay would already be working in multiple languages when they speak in Spanish. Victoriano chastises the Kumeyaay for their taunts in Spanish, and readers are supposed to appreciate his perspective, especially as his words lead directly to action: "This rebuke and imperative order silenced them immediately, and not understanding why these gentlemen were having all that fun, and did not laugh, nor wished any one else to laugh, quietly turned and went home."[34] In an almost farcical scene, though with darker undertones, the issue of *The Squatter and the Don* is seen in miniature here, and with almost all the races and ethnicities represented (excepting Mrs. Darrell's African American maid, Tisha). The Kumeyaay see something and interpret it in one way and are scolded for their misreading, without anyone taking the time to make the situation clear to them. Szeghi observes: "By assuming that another person might be treated like an animal for sport, the vaqueros position themselves closer to William Darrell than to the Alamars" in Victoriano's reading of them; he scolds them after assuming that "Indian vaqueros are incapable of properly reading a situation."[35] She goes on to criticize Ruiz de Burton for making her Kumeyaay characters "ontologically ill equipped" for mental labor: "Their 'perfect' execution of manual labor contrasts with their limited understanding of more complex human interactions."[36] I agree with Szeghi's analysis of Ruiz de Burton's attitudes, especially in terms of Victoriano's reading of the Kumeyaay characters, though I would add that there is another, most positive, reading possible. The Kumeyaay in fact have read the situation correctly and influence the Alamars and Darrells through their actions.

Soon after the initial action, Victoriano and Everett—Darrell's son—begin to laugh from the spectacle as well. They are also scolded for their behavior, this time by Gabriel—but not as strongly as the vaqueros had been earlier. Gabriel merely says, "For shame, Victoriano, to be so discourteous." Ruiz de Burton adds, though, "But Victoriano had suppressed his desire to laugh too long, and now his risibility was beyond control. Everett was overcome in the same manner, so that he hung on Victoriano's shoulder, shaking with ill-suppressed laughter."[37] In other words, the Kumeyaay should not have been blamed for their reaction. They have read the ridiculous scene in the same manner as the primary characters. Moreover, they have showed their vaquero skills at the same time, correcting the Californio's technique. Indeed, their suggestions may have been what truly spurred the harsh comment—out of Victoriano's jealousy of their technical knowledge, perhaps—rather than an abundance of concern for etiquette. Moreover, in this same scene, Gabriel Alamar calls William Darrell a "wild cattle" who must be prevented from "striking my father." He continues: "I would *lazo* you again fifty times, or any other man, under the same circumstances. If you think it was cowardly to do so, I will prove to you at any time that I was not prompted by cowardice."[38] At this point, everyone is being compared to animals, and everyone is reading the action around them questionably. Kumeyaay, Californios, and squatters display their vaquero skills—or lack thereof—in this passage where people are handled as cattle and then judged for their cattle-based skills.

In this scene, Ruiz de Burton challenges stereotypes of civilization through regionalism. Susan Gillman compares the regionalist trend in U.S. literature and culture in the second half of the nineteenth century, focusing on the Southwest and the U.S. South. She reads José Martí's theory of a *mestizaje* in the work of Harriet Beecher Stowe, Helen Hunt Jackson, and Ruiz de Burton. Martí links Stowe and Jackson, the latter of whose *Ramona* was published a year before *The Squatter and the Don* and similarly dealt with California, Native American characters, and U.S. westward expansion. *The Squatter and the Don* and these other key texts can be read hemispherically as a palimpsest of what Martí calls "our mestizo America." Ruiz de Burton's novel represents Martí's *mestizaje*, a transcultural and transnational California seen in her complicated portrayal of the intersections between U.S., Californio/a, and Kumeyaay characters.[39]

The image of the vaquero in the West details the sentimentalism at work in the genre. Karen Kilcup notes that Ruiz de Burton's novel "appeals to logic, emotion, and ethics" as she works to convince her U.S. elite readership to her Californio/a cause.[40] Ruiz de Burton uses sentimentalism strategically in her writing, following the style of Stowe, who famously paired her Euro-American and African American characters in *Uncle Tom's Cabin* (1852). Stowe gave African American and Euro-American characters the same names—two Georges, for example—linking characters, regardless of race, through their paralleled names. Ruiz de Burton interprets this rhetorical technique by contrasting people and agricultural methods, instead. The squatters contrast the Don, while wheat farms (the U.S. squatters) contrast cattle ranches, vineyards, and fruit trees (Don Mariano). Note that the element of pairing remains in both works. David Luis-Brown calls *The Squatter and the Don* a "reversal of Stowe's vision" focusing on "landed white Southerners and elite *Californios*."[41] Ruiz de Burton works to show that following the Treaty of Guadalupe Hidalgo these Californios/as would be injured by unfair laws brought with U.S. settler colonialism. For Ruiz de Burton, contrasting sentimentalized peoples and agricultural methods proves to be an effective way to engage her readers' emotions: outrage, insult, and humor. She uses this technique for the Alamars in their fight against the U.S. squatters, while the Kumeyaay remain in the middle ground.[42]

Ruiz de Burton uses her understanding of rhetoric to make a strong case for Californio/a rights. Don Mariano says early on: "While we are waiting to have my title settled, the *settlers* (I don't mean to make puns), are killing my cattle by the hundred head, and I cannot stop them."[43] Interestingly, his skill (and that of Ruiz de Burton) with English as his second language comes into great account in this scene. He shows himself to have enough mastery of his second language to be able to make puns, like the novelist herself. In other words, Ruiz de Burton, as seen in her protagonist, is in control of her language in *The Squatter and the Don*. She critically reads various legal documents while establishing the cattle-related portion of the conflict, inviting her readers to engage with her language closely in these scenes as well. This close attention to language makes her rhetorical choices potentially rich sites for study, and her use of describing people as beasts becomes a significant choice. Even Clarence Darrell, the U.S. capitalist who makes a fortune and saves the Alamars by marrying Mercedes at the end of the novel, falls into the habit of discussing people as beasts.[44] When he speaks with Doña Josefa Alamar about buying all of Don Mariano's cattle to feed miners at his property in Arizona, he concludes: "Don Mariano can buy all the cattle he wants to re-stock his rancho after he gets rid of the two-legged animals."[45] After the Darrells and Alamars nearly come to blows, treating each other like cattle, the novel's heroic figure shows a similar inclination. Once again, the scene that confuses most of the main characters of the novel, but which the Kumeyaay have truly understood, perhaps also influences Clarence Darrell's word choice. While many readings of the novel focus on the contrasted Californios/as and squatters, the Kumeyaay characters are also represented in ways that complicate this reading. Ruiz de Burton does so by having multiple characters act out escalating hostilities through dehumanization in a scene out-of-doors on the ranchero, directly in the agricultural environment. At the same time, she shows the skills of her Kumeyaay characters, their correct ability to understand complicated cultural power dynamics, the harsh rebuke they receive for that correct—if unwelcome—understanding, and ultimately their impact on the other major characters in the novel, who follow their course of action with their own laughter. Unnamed and largely ignored, the Kumeyaay demonstrate their resilience and influence in *The Squatter and the Don* through key scenes like this one.

Vaqueros in the Storm

The other significant scene connected to cattle in the novel is even more directly connected to the natural world: a cattle drive through the Sierra Nevada interrupted by a snowstorm that leaves Don Mariano and Victoriano Alamar severely injured and their cattle presumed dead. In a novel that does not rely much on physical description, Ruiz de Burton in this scene provides sensory details to add to the drama of the snowstorm in the mountains:

> The shouts of the *vaqueros*, bellowing of cattle and barking of dogs resounded throughout the valley, the echo repeating

them from hill to hill and mountain side. In a short time everything living was in motion, and the peaceful little valley seemed the battle-ground where a fiercely contested, hand-to-hand fight was raging. The great number of fires burning under the shelter of trees, seen through the falling snow as if behind a thick, mysterious veil, gave to the scene a weird appearance of unreality which the shouts of men, bellowing of cattle and barking of dogs did not dispel. It all seemed like a phantom battle of ghostly warriors or enchanted knights evoked in a magic valley, all of which must disappear with the first rays of day.[46]

Ruiz de Burton challenges notions of California scenery, representing it not as idyllic but rather dangerous. As opposed to a scene of cattle being driven across the state to Arizona mines, she describes this action as a "hand-to-hand fight." This sense of struggle is complemented by the campfires dotted along the mountain and the sounds of men shouting. Readers experience the spectacle through multiple sensory details. Ruiz de Burton takes this metaphor so far as to call the vaqueros "ghostly warriors or enchanted knights." Her sentimental rhetoric appears in the idea of this as a premodern battle against the elements rather than against a human opponent. The novel is focused on visualizing sentiment that has to do with Californio/a economics rather than landscape vistas. Yet this scene does, truly, set a scene. Words like "ghostly" and "enchanted" emphasize Ruiz de Burton's rhetorical flourishes. She is not focusing on idyllic realism in this scene; rather, she creates a sense of the romantic with its setting in the Sierra Nevada and the strangely falling snow.[47]

As often occurs in Ruiz de Burton's novel, the Kumeyaay characters are not discussed in detail, though they are shown to be working throughout the scene. "The *mayordomo*, with about twenty *vaqueros*, were nearly at the foot of the mountains with twenty-five hundred head of cattle, when Don Mariano and Victoriano overtook them, and as the cattle had been resting there for two days, their journey to the Colorado River would be resumed at daybreak."[48] The mayordomo (foreman) in this scene could be a Kumeyaay character, potentially showing job mobility for these largely anonymous workers. He makes a report to Don Mariano that is given in English in the novel, though presumably it would have been spoken in the Spanish that both characters would have in common. "With a good supper and good night's rest, we will make a long march to-morrow," he says, adding, "There is plenty of feed here for our cattle."[49] The mayordomo's assessment proves tragically wrong, though, as a brutal snowstorm kills the cattle and almost kills the Alamars. Ruiz de Burton writes: "The snow had not ceased falling for one moment, and if the *mayordomo* had not been so good a guide they might not have found their way out, for every trail was completely obliterated, and no landmarks could be seen."[50] Once again, Ruiz de Burton almost unintentionally compliments her Kumeyaay characters. They are able to navigate better than her Californios, who are in turn considered superior to the U.S. squatters. In the vaquero culture it is important to be able to drive cattle efficiently and correctly across great distances. The Alamars become hopelessly lost in a snowstorm, an unfortunate but not impossible eventuality. The Kumeyaay, however, are able to at least save the Californios from their inability to navigate—though this might be what makes them unable to save the cattle, which are seemingly lost in the storm.

The vaqueros have already had to stay up all night with the cattle to keep them from freezing to death. They have to "whip up and rouse the cattle" throughout, fueled only by coffee and their own adrenaline. At this point, the cattle are healthy, though they will seemingly die from exposure as the storm continues. However, Don Mariano and Victoriano—who had only arrived the evening before—are both soon nearly perishing from the cold. The vaqueros abandon the herd to rescue the Alamars, returning them to safety. After their ordeal, Don Mariano "had taken a severe cold in his lungs" that kept him in bed "for six weeks."[51] Meanwhile, Victoriano becomes paralyzed in his legs from the cold, a condition that will remain with him for the rest of his life. Ironically, the fact that Ruiz de Burton shows her Kumeyaay characters to be robust can be seen as a further sign of her racism, as delicate constitutions were in vogue during the Victorian period, and the frailty of the Alamar men reveals their elite status. Elite Euro-American women of the Victorian era often suffered from "nervous complaints" (Mercedes Alamar faints repeatedly throughout the novel). More interestingly, though, even Ruiz de Burton's male Californio characters are described as physically delicate. These are not strong laborers, but creatures that grow frail—even to the point of death—when they are pushed to any physical extreme. This proves their elite, "European" status when compared to the hardy laboring Kumeyaay. At another point in the novel, the character Mechlin lauds the "salubrious air of San Diego" for its healthfulness. His health had "seemed permanently undermined," and he appeared to be a "living skeleton" before California's climate restored him to health.[52] While Ruiz de Burton puts the language of this health tourism in the squatters' mouths, perhaps these health concerns really should have been attributed to the delicate Californio/a characters, who seem to need the restorative properties attributed to the region. The ailments they receive during the snowstorm leave them with lasting, fatal damage. Moreover, the true survivors are proven to be the Kumeyaay and other native peoples of the region, as they have historically suffered from Euro-American disease and ill treatment, and recently through the same snowstorm that devastated the Alamar men. Ruiz de Burton's concerns with health in the novel can be seen in this instance in which the vigorous

Kumeyaay outlive even the doomed cattle. Their robustness can be read as an element of dehumanization that adds to Ruiz de Burton's racial stratification of the Californios/as, Kumeyaay, and even cattle. But, however Ruiz de Burton defines status, the fact remains that the Kumeyaay characters save the Alamars in this storm scene, themselves presumably surviving, demonstrating another element of resilience against centuries of threats to their health and lives.[53]

The storm is discussed once again at a meeting between Don Mariano and the railroad magnate and former California governor Leland Stanford on behalf of a group of Californio/a investors hoping to form a southern railroad connecting San Diego with the southern Atlantic seaboard, complementing the northern railroad controlled by Stanford and his colleagues. Brook Thomas discusses the novel and its railroad plot. The southern railroad speculation was an issue during the second half of the nineteenth century: "Documented by W. E. B. Du Bois, that support is also recorded in the positive role railroads play in pro-Reconstruction novels by Albion W. Tourgée and Charles W. Chesnutt."[54] The railroad is ultimately doomed due to the northern railroad monopolists, including Stanford, who do not want the competition and make bribes to influence policy. Thomas connects the results of this monopolization of the nation's expanding cattle industry and the railroad in a way that can also be seen in Chesnutt's writing and would come to ultimate literary fruition in Upton Sinclair's *The Jungle* (1906). Don Mariano wants this industrialization of cattle to bring wealth to the Californios/as, as well. After Stanford asks if he has cattle, the Don answers, with some spirit, "No, sir, I haven't. The squatters at my rancho shot and killed my cattle, so that I was obliged to send off those that I had left, and in doing this a snow-storm overtook us, and nearly all my animals perished then. The Indians will finish those which survived the snow."[55] To this, Stanford asks, "Those Indians are great thieves,

I suppose?," allowing Don Mariano to reply, "Yes, sir; but not so bad to me as the squatters. The Indians kill my cattle to eat them, whereas the squatters did so to ruin me."[56] The Don realizes that there are appropriate and inappropriate ways to deal with the land and livestock. The Kumeyaay are preferable to the squatters in their motivation for killing cattle. Richard Slatta adds a real-world example to corroborate Don Mariano's statement: "I could describe the attitude of cattlemen toward rustlers, but why not let Colonel E. P. Hardesty, an Elko County, Nevada rancher, speak for himself? He instructed his cowboys how to handle a man who stole any of his cattle: 'If he stole it to eat, tell him to enjoy it and bring me the hide. If he stole it to sell, bring me his hide.'"[57] In either case, there is a cowboy code whereby it is better to kill to eat than to waste or use in some other manner. Don Mariano appreciates the fact that the Kumeyaay honor this code in a manner that the squatters do not.[58]

Ruiz de Burton makes one of her most dramatic points about the injustices against the Californios/as during an actual storm. The wilderness as a symbol comes about in large part because of the Treaty of Guadalupe Hidalgo, as vast expanses of new land gave the United States the opportunity to erase the Mexican Americans from their former lands in much the same way as they had with Native Americans before them—making those lands open to new uses. Moreover, Ruiz de Burton plays on the "cattle-thief" stereotype as seen in Winnemucca's work, where U.S. squatters punish the Paiutes for acting out of hunger, but in this instance Don Mariano shows sympathy to the Kumeyaay. He argues that at least they kill the cattle to eat them; on the other hand, the squatters kill the cattle to let them rot, not unlike those who had killed the bison en masse elsewhere in the U.S. West. In a way that can be compared to the paralleling of humans, animals, and the natural environment, Ruiz de Burton's argument is not unlike Henry David Thoreau's call for "Higher

Laws" as opposed to the people's laws, though Thoreau is one of those New Englanders Ruiz de Burton would have mocked in her earlier novel, *Who Would Have Thought It?* (1872).[59] Szeghi observes that by this point Ruiz de Burton might have learned to at least present a show of sympathy with Native American characters after all: "Don Mariano's suggestion of affinity with Indians can be read as a calculated political move meant to distance him from Anglo American political corruption and place him on the side of its victims."[60] The Don is either being strategic or is honestly sympathetic, learning to empathize with those who have suffered in the face of his own suffering. Throughout, Ruiz de Burton writes about a specific moment in the history of California, transitioning from Mexico to the United States, and her views and attitudes about race, gender, and cattle ranching add to her argument in favor of land sovereignty for Californios/as. In this reading of *The Squatter and the Don*, she describes her Kumeyaay characters as the most capable vaqueros. In an extension of this reading, it is notable that perhaps the most skilled vaquero is Chapo, the only named Kumeyaay character, but one whose actions harm the main characters.[61]

Chapo: The Individual Vaquero

The case of Chapo is one of the most damning in terms of Ruiz de Burton's attitude toward her Kumeyaay characters. In a key scene, Chapo is called on to perform his vaquero work: "'Yes, *patroncito*, I'll do it right away,' said the lazy Indian, who first had to stretch himself and yawn several times, then hunt up tobacco and cigarette paper, and smoke his cigarette. This done, he, having had a heavy supper, shuffled lazily to the front of the house, as Clarence was driving down the hill for the second time."[62] Ruiz de Burton describes Chapo as having no intention of performing his work efficiently. In the time he wastes, the Alamars and Darrells have missed an opportunity to resolve their conflict, and this missed chance leads to a series of tragic events. Indeed, this scene is set up as a causal factor to the tragic endings of several characters in the novel, the Alamars in particular. Peter Chvany, among others, has observed that although Chapo is the only named Native American character in the novel, his treatment is perhaps the "most disturbing of all": "When performing their labors well, as when they help the Alamares save cattle from a blizzard, they are deferentially silent."[63] Therefore, when they stand out they do so for negative reasons. They are interfering with the lives of Californio/a characters and are to blame for the novel's tragedies. I agree that Ruiz de Burton's treatment of Chapo is disturbing, and I sympathize with Chvany's argument that "rather than faulting *The Squatter and the Don* for participating in the very racism it seeks to overturn, we can understand its 'blind spots' productively."[64] Certainly, I would argue that her "blind spot" about her vaqueros is a rich site for productive readings of the novel. However, what if we make the effort to read Chapo's actions as resistance? His acts have great power in this scene, comparable to those of Leland Stanford. In that way alone, Ruiz de Burton shows the power of the Kumeyaay, even while disparaging them.[65]

While this scene is often used to show Ruiz de Burton's racism against her Native American characters, her racism is in the passive sense, not the active. This reading is rhetorically framed in her use of the word "lazy" twice in one passage. The term is used only a few other times in the novel: in a teasing manner when Mrs. Mechlin urges her partner to dance, and when Don Mariano describes what he imagines the squatters think of the Californios/as: "Then the cry was raised that our land grants were too large; that a few lazy, thriftless, ignorant natives, holding such large tracts of land, would be a hindrance to the prosperity of the State, because such lazy people would never cultivate their lands, and were even too sluggish to sell them. . . . The settlers want the lands of the lazy, the thriftless Spaniards."[66] To Don Mariano, the term "native" refers to Californios/as; he does not use this term for Native American characters. Also, he emphasizes the term "lazy" even more in this paragraph than the narrator does with Chapo. Don Mariano's melodramatic reading is supported later by the squatter Mathews, who in fact admits to this belief: "These Californians are too ignorant to know how to defend their rights, and too lazy to try, unless some American prompts them."[67] Perhaps despite herself, then, Ruiz de Burton links the Kumeyaay with the Californios/as through the term "lazy," which is applied to them in each case (imagined or in reality) by others. Both cultures are othered in respect to their work ethic with that same term. Ruiz de Burton parallels the treatment of the Californios/as by the U.S. squatters to the treatment of the Kumeyaay by the Californios/as themselves. Perhaps Chapo's laziness is something else, or misread, or a mere coincidence. If we as readers are not supposed to believe the words of the squatters, having learned to sympathize with the Alamars, then perhaps we should also distrust the same word being applied to the Kumeyaay. We have to some extent been trained that way through the rhetoric of the novel.[68]

One of the final representations of Kumeyaay in *The Squatter and the Don* places them in a scene in which they act alongside the U.S. squatters themselves, a rarity in the novel. Usually, they work as vaqueros with the Californios/as, not with the squatters who grow wheat. In this scene, though, Ruiz de Burton sympathizes entirely with the Kumeyaay characters against another hypocritical squatter. In this case, John Gasbang, now an upstanding member of the community and a religious leader, is shown to have gained a significant portion of his wealth by gambling, playing "monte" with the Kumeyaay: "Pitilessly would John strip his unsophisticated tattooed comrades of everything they owned on this earth. Their reed baskets, bows and arrows, strings of beads, tufts of feather-tips, or any other rustic and barbaric ornaments. All, all, John would gather up with his skillfully shuffled cards."[69] While Ruiz de Burton might not entirely respect her Kumeyaay characters for more or less allowing themselves to be duped, her bitterness is decidedly against the squatter and his community. She continues, ironically, "Kindly San Diego had forgiven John's petty thieving. The money won from the poor Indians had helped him to thrive, and consequently convinced him that, after all, cheating was no worse than other sins, the gravity of which entirely depended upon the trick of hiding them."[70] Her sarcasm drips in terms like "kindly" and "thrive." Gasbang's determination that his failings are "no worse than other sins" is starkly contrasted to the idea of the Kumeyaay losing "everything they owned on this earth." The addition of the phrase "on this earth" adds to the sympathy Ruiz de Burton causes her readers to feel for the victims of Gasbang's practices, her ongoing theme of land use and rights. In this, Ruiz de Burton again cannot seem to help herself with showing sympathy to the Kumeyaay. They are being cheated by the U.S. squatters in a parallel fashion to the Californios/as themselves. Indeed, they are similarly losing everything they own "on this earth" to the U.S. squatters and railroad monopolists. Rather than the clichéd "beads and feathers" Ruiz de Burton lists for the Kumeyaay possessions, the Alamars lose their land and their cattle. She describes a different scale of possession, but they are representative of great loss in both instances: both groups lose "everything they owned on this earth," the Kumeyaay people for the second time—after colonization. Gasbang's treatment of the Kumeyaay—cheating them in gambling, for example—demonstrates his hypocrisy and classlessness that Ruiz de Burton has been establishing throughout the novel. The squatters are thieves, and in this case, they perform on a smaller scale the theft that they, and later the railroad, have acted out against the Californios/as. Ruiz de Burton places Californios/as and Kumeyaay on a level, then, connected by their great loss.

Conclusion: Vaquero Downfall

At the end of the novel, the squatters win, the Alamars lose their land, cattle, and health, and California becomes a wheat-growing state. Alicia Contreras notes that "California would, to an even greater extent, turn away from the ranching and selective farming advocated by Don Mariano and engage in a series of destructive water wars, as depicted in Mary Austin's regionalist novel *The Ford* (1917)"—and, I would add, the railroad wars depicted in Frank Norris's *The Octopus* (1901).[71] Ruiz de Burton ends her novel with one of the most clichéd finales in U.S. narrative since the union of John Rolfe and Pocahontas: a marriage between the "exotic" ethnically marked woman and the moneyed U.S. American man. A protofeminist herself, Ruiz de Burton's main female character, Mercedes Alamar, is not as strong, or as protected, as her creator, conforming to Victorian stereotypes. Ruiz de Burton was better prepared for her politicized and economic life than many of her contemporary U.S. women authors. It is because of these contradictions, though, that Ruiz de Burton becomes a fascinating study for a particular period in U.S. literature and culture—specifically the ways in which she engages with cattle and Kumeyaay vaqueros during key scenes of her novel.[72]

In their influential introduction to the 1992 republication of *The Squatter and the Don*, Rosaura Sánchez and Beatrice Pita observe that "her novel would be the first published narrative written in English from the perspective of the conquered Mexican," adding that the novel "goes beyond this binary opposition" of squatter and don merely—that it "calls for a double reading."[73] I agree; the novel is able to be read more richly than Ruiz de Burton initially seems to intend. As editors of her *Letters*, Sánchez and Pita also note that the novel is a "case study of an individual who emerged from the lot of a conquered people with a clear, if fissured, racial memory, that is, with a consciousness of *latinidad* regarding what she perceived as Anglo cultural, political, and economic hegemony."[74] In addition, doubling seems to be an important technique for Ruiz de Burton. Alemán reads the novel as a form of Bakhtinian double-voiced discourse. Ruiz de Burton is excellent at rhetoric, yet she does not appreciate her own hypocrisy. She uses techniques in her writing that would seem to encourage her readers to rethink prejudices against Kumeyaay characters, but she does not question her own prejudice. In addition, she argues strongly for her Mexican American people but speaks directly to a Euro-American audience. Ana Castillo writes that Ruiz de Burton's novel is an "unmitigated call to arms. . . . In the United States, the movement of Manifest Destiny that led to war with Mexico is recorded as the Mexican-American War. In Mexico, it is regarded as the 'U.S. Invasion.' This ideology was, and, it may be argued, remains the premise for this most powerful nation."[75] Ruiz de Burton's argument was not successful upon publication, largely ignored even by readers of western regionalism. The movement of Euro-Americans and their cattle continued inexorably across the continent throughout the rest of the nineteenth century. Even when cattle were already present, U.S. settlers challenged the status quo in a way that destroyed the established cultures in place. The Mexican cattle, like the Mexican and Kumeyaay people, were largely removed from California wealth and significance, replaced with U.S.-controlled farms and their domestic plants and animals. Of course, cattle would remain in California, as cattle were being raised throughout the U.S. West, seen in a 1938 photograph of cattle in the California hills (see fig. 13).[76]

This change created lasting consequences in California and national culture. Although in many ways a difficult novel, *The*

Fig. 13. In this 1938 photograph by Dorothea Lange, cattle are feeding in the California hills. Although this photograph comes long after the events depicted in *The Squatter and the Don*, the image is similar to what Ruiz de Burton would have known on her land at Rancho Jamul. Library of Congress, Prints & Photographs Division, Farm Security Administration, Office of War Information Black-and-White Negatives.

Squatter and the Don is useful to show the reactions by one community to the encroaching tide of U.S. settlers, their farms, and
their livestock. Ybarra notes: "Unfortunately, in one of the book's
greatest ironies, don Mariano's plan fails because the Americans
cannot imagine themselves as vaqueros, or cowboys—an image
that these days stands as one of the globally most recognizable
symbols of the American West."[77] Again, this is an ironic conclusion, as cattle would later be again successfully raised throughout
the U.S. West over the twentieth century. Alemán sums up: "To
paraphrase the title of Ruiz de Burton's first novel, who would
have thought a nineteenth-century Hispana would publish for an
Anglo readership a narrative in English that exposes the racism,
class oppression, and dubious legal maneuverings that underlie
Anglo America's romantic conception of the nation's history?"[78]
Ruiz de Burton's novel makes a complementary reading to Sarah
Winnemucca Hopkins's work. In both cases, Euro-Americans
have effected dramatic change on the people of the West; also,
both engage Native American characters, Paiute and Kumeyaay,
though to different effect; finally, both use cattle as significant
symbols for their arguments about land, sovereignty, and food
justice. In the following chapter, Charles Chesnutt in fact does
not use cattle as a significant symbol in his *Conjure Stories*, which
are set in post-Reconstruction North Carolina. As U.S. territorial
expansion focused on the West in the late nineteenth century, the
South worked to renegotiate the configurations of farms, labor,
and livestock following the failure of Reconstruction. Pigs and
chickens became the dominant livestock animals in the region
during this time, much as they are in the twenty-first century. I
argue that this move—the relative absence of the nation's dominant animal in the South—is significant itself in the negotiation
of power in terms of race, the environment, and the U.S. farmer.

SIX

Southern Cuisine without Cattle in Charles Chesnutt's *Conjure Stories*

In 1899 Charles Waddell Chesnutt published his first full-length work, *The Conjure Woman*, the book that would become his most successful publication during his lifetime (see fig. 14). This series of regionalist stories set in late-nineteenth-century North Carolina centers on three main characters: Julius, sometimes referred to as Julius McAdoo or Uncle Julius, an emancipated man who lives on land recently purchased by two northerners, John and Annie.[1] The three now live in close proximity to each other, and Julius frequently visits the northern couple, telling them stories of his antebellum experiences in the neighborhood to amuse them and to benefit himself. Julius's tales are framed by John's narration, and together they tell the cultural story of the region. Julius describes the ways that antebellum African Americans navigated their lives within the system of slavery, often with the assistance of conjure—a cross between magic and folk knowledge. In the voice of Julius, Chesnutt applies the dialect writing popular in late-nineteenth-century U.S. regionalism. He also engages with the theme of regional African American cuisine in the South through his descriptions of pig—largely ham and bacon—and chicken in the tales. Chesnutt wrote fourteen *Conjure Stories* overall, only seven of which were selected for publication in *The*

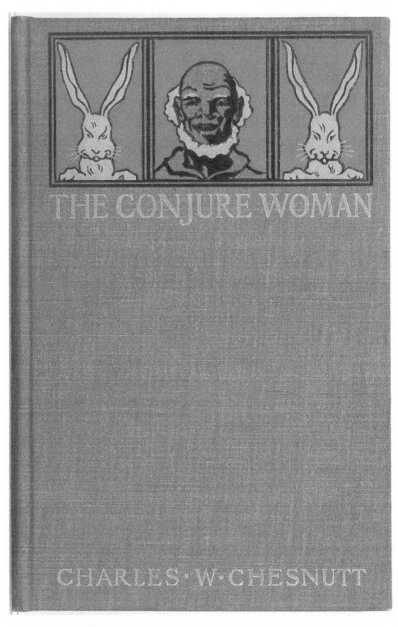

Fig. 14. Cover of Charles W. Chesnutt's *The Conjure Woman*, 1899. Note that it has been designed to be comparable to Joel Chandler Harris's "Br'er Rabbit" works. Collection of the Smithsonian National Museum of African American History and Culture.

Conjure Woman; I will discuss other *Conjure Stories* in this chapter as well. Throughout, the role of animal husbandry as practiced in the South is as significant in the stories as is Julius's dialect. For example, in the tale in "Mars Jeems's Nightmare," one of the characters gives an agricultural survey: "Solomon up en tol' 'im 'bout de craps, en 'bout de hosses en de mules, en 'bout de cows en de hawgs" (Solomon up and told him about the crops, and about the horses and the mules, and about the cows and the hogs).[2] Solomon's list appears to have a rank and order: crops are first, but unspecific; horses and mules—laboring animals— come next; finally, the cows and hogs, animals raised for meat, are described—perhaps in order of symbolic value. In this way, Chesnutt shows the agricultural attitudes in the South during the second half of the nineteenth century. These attitudes—especially as concerning pigs and chickens—make a striking contrast to those I have described elsewhere in *Cattle Country* regarding cattle as the dominant animal for the nation as part of nineteenth-century agri-expansion. In the *Conjure Stories,* pigs and chickens become the iconic livestock, foundational for the region's distinct culinary culture. In addition to his critiques of racism throughout his stories, then, Chesnutt also celebrates the agricultural and culinary significance of the South both preceding and following the Civil War with these key symbolic animals.[3]

At the end of the nineteenth century, agri-expansion and the closure of the frontier had fundamentally changed the physical layout of the United States, and perhaps in no industry was this as pronounced as in agriculture. As most of *Cattle Country* discusses the movement of cattle west with Euro-American settlement as well as the reactions to this drive in nineteenth-century U.S. literature and culture, this chapter engages with a region that represented a major gap in the nation's cattle-based agri-expansion: the South. After the Civil War, the largest form of agriculture in the southern states—plantation agriculture dependent on unpaid human labor—had indeed changed. The Civil War caused a fundamental shift in U.S. economics, from conquering wilderness to developing industries. These industries were growing to include cattle-based monoculture across the West, as well as intensive pig and chicken production in the South. In his *Conjure Stories*, Chesnutt focuses on much southern regional culture—seen in his use of dialect writing and the inclusion of conjure. He presents similarly focused readings of southern agriculture and cuisine in his stories, as well. His use of dialect writing and conjure as regionalist elements has been discussed elsewhere, but I argue that his use of food items in his stories is equally significant. I specifically examine the contrast between cattle—the subjects of scenes in only two of the *Conjure Stories*—and the more dominant pigs and chickens, thereby comparing some of the country's most produced meat animals. Pigs and chickens, while popular foods throughout the United States, have been largely connected to the South culturally rather than to the nation at large. Even in the two stories in which Chesnutt mentions beef in any detail—namely, "The Conjurer's Revenge" and "A Victim of Heredity: Or, Why the Darkey Loves Chicken"—he still privileges pigs and chickens. Choosing to do this in *Conjure Stories*, Chesnutt shows a kind of regional culinary and agricultural pride. As in the work of Henry David Thoreau considered earlier, Chesnutt describes the prominence of regionally specific foods to challenge the national dominance of cattle. I weave together elements of Chesnutt's writing—race, land, and U.S. animal husbandry—as they are used in these two *Conjure Stories* specifically in order to show Chesnutt's commentary on a culture that practices racism as well as environmental degradation in its mode of agriculture, while also praising the culinary element of southern culture as a contrast to the nation's cattle-connected agri-expansion.[4]

Chesnutt understood the nation's cultural drive to produce and consume beef as well as anyone during that time. He had firsthand experience with the agricultural life of sharecroppers during and after Reconstruction in North Carolina. His father, Andrew Jackson Chesnutt, worked as a small-scale farmer after the decline of the Chesnutt family store, which corresponded with the time of the failure of Reconstruction in the area. As cattle had become a dominant image for westward expansion in the United States, chickens and pigs came in part to represent a significant element of southern agriculture, both before and after the Civil War. While southern farmers attempted to raise cattle well into the twentieth century, they had far greater success in producing large quantities of pigs and chickens. This distribution of livestock is maintained in the popular culture as well. Around the same time that Chesnutt was writing his *Conjure Stories*, in 1895, Blackwell's Durham Tobacco Company created an advertisement that used an image of the representative North Carolina "Farm Yard." Pigs and, especially, chickens are dominant in the image, one that comes complete with peacocks but noticeably without cattle (see fig. 15). Furthermore, Chesnutt shows his knowledge of agri-expansion in a brief scene from his non-conjure 1899 story, "Uncle Wellington's Wives." In this story, the title character notes upon entering the yard of a particularly grand house in Ohio: "I don't like dem big lawns. It's too much trouble ter keep de grass down. One er dem lawns is big enough to pasture a couple er cows" (I don't like them big lawns. It's too much trouble to keep the grass down. One of them lawns is big enough to pasture a couple of cows).[5] In this scene, Chesnutt is not presenting anyone becoming a cow, for example. This is not one of his conjure stories. Rather, the character is considering the economics and agricultural practicality of lawns in a wealthy Ohio neighborhood. To Wellington, lawns look much like pasturelands for cattle, but even a very large lawn can only maintain one or two cows—they are excessively demanding on the land needed to raise them. Rather than the South, the West and Midwest are marked as appropriate spaces for cattle, even in this brief instance in one of Chesnutt's non-conjure stories.

In Chesnutt's time, pork was the food item most available to the largest number of U.S. Americans, one that could be eaten whenever the preferred beef was not available.[6] As far back as British colonization of the lands that would become the U.S. South, pork and chicken were more common than beef. On the southern farm, "other than a couple of chickens scratching in the yard and perhaps a wandering pig or two, scarcely any livestock would be visible at all," Virginia DeJohn Anderson observes.[7] Indeed, the former European colonists living in the regions of Virginia and the Carolinas quickly moved beyond trying to maintain the land by the mid-seventeenth century, instead turning their attention to cash crops like tobacco and cotton. Planters continuously returned to mining new lands as opposed to sustaining the fields already under cultivation. Older, worn soils were difficult to work, whereas fresh soils were highly fertile, encouraging an exploitative mentality exacerbated during the plantation slavery system. Landowners treated the soil itself as a commodity, not considering the matter of sustainability. This resulted in a diminishing return of yields that strained those who owned and as well as those who worked on the land—largely an enslaved population. The underlying attitude in the Carolinas, then, according to Steven Stoll, was in "spending fertility when cotton was high no matter the long-term consequences."[8]

Far from repairing that damage, U.S. farmers continued their search for fresh soils west. Interestingly, Stoll wonders if some of the agricultural developments at the national level, such as agri-expansion, did not get their start with this wasteful ecology: "Did improvement draw the starting line for the technological scramble that led to industrialized agriculture?"[9] The soils of the South being worn, and unpaid laborers no longer an option, the nation's agri-expansion in terms of intensive cattle production

Fig. 15. This advertisement from the Blackwell's Durham Tobacco
Co., printed in 1895, around the same time as Charles Chesnutt's
The Conjure Woman, depicts the animals of a North Carolina
farmyard. Note the absence of cattle. Library of Congress, Prints
& Photographs Division, Marian S. Carson Collection.

in particular moved away from the South toward the West. The
southern region continued to work with animals that were less
demanding on the depleted soils. North Carolina has continued
its connection to chickens and pigs into the twenty-first century.
According to United States Department of Agriculture (USDA)
reports, while the state produces negligible amounts of cattle for
slaughter, it ranks second in the United States in terms of industrial pig production and third for broiler-type chickens.[10] Forests

were cleared for grazing lands for cattle in other regions of the expanding nation during the nineteenth century; however, such was not the case in the U.S. South. In addition to the continued predominance of cash crops like cotton and tobacco, and occasionally specialty crops as in the example of John's vineyard in the *Conjure Stories*, the food animals that ultimately became identified with the region's cuisine and its culture were pigs and chickens.[11]

The popularity of pigs and chickens in stories about the South has resulted in stereotypes of southern African Americans as "ham-stealers" and "chicken lovers" as part of a larger racist U.S. cultural history. Psyche Williams-Forson analyzes minstrel-style materials that play on such stereotypes popular in the nineteenth century.[12] In addition, Kyla Wazana Tompkins reads a Magnolia Hams ad from 1878—shortly after the inauguration of Rutherford B. Hayes, the official failure of Reconstruction, and the introduction of Jim Crow laws to the South—in which caricatures of African American characters are seen eating the named ham.[13] While pigs and chickens were more economical forms of animal protein, their stereotypical associations with African American culture became disproportionate compared to other protein sources. They became fraught food animals, therefore, in African American literature as well. Throughout the *Conjure Stories*, then, Chesnutt challenges these stereotypes and racism in U.S. society, working to reclaim key elements of southern African American culture such as its cuisine and agriculture.[14]

Like Chesnutt, Booker T. Washington and W. E. B. Du Bois, two of the most influential cultural critics of the late nineteenth and early twentieth centuries, also analyzed the comparative value of beef, ham, and chicken in their writings. Beef, relatively rare in the South in general, was of great cultural significance at Washington's Tuskegee Institute, for example.[15] For Washington and the Tuskegee cooks, diet was a way to build new habits and reject those marked regionally in a negative way. Tuskegee served "roast beef, veal cutlets, beef gravy, soup, and liver and onions," writes Jennifer Jensen Wallach; "Washington and other members of the Tuskegee staff shared some of the culinary prejudices of self-consciously refined African American eaters who wished to disassociate themselves from foods like pork that were evocative of slavery or southern regionalism."[16] Pork and chicken became negatively associated in this way; beef, on the other hand was viewed as being more civilized. In addition, beef was harder to preserve than pork was, so beef remained a more difficult meat to obtain—and therefore of higher status—into the twentieth century. Washington notes in his autobiography that while the Tuskegee Institute kept "over two hundred horses, colts, mules, cows, calves, and oxen," that number was overwhelmed by "about seven hundred hogs and pigs."[17] While he does not give the number of chickens, it is clear that the hogs and pigs dramatically outnumbered all other four-legged animals, despite attempts to privilege beef for its assimilating effects. Meanwhile, Du Bois also focused on "the deceitful Pork chops [that] must be dethroned in the South and yield a part of its sway to vegetables, fruits, and fish" in his 1918 article, "Food," in The Crisis.[18] Chesnutt's Conjure Stories provide another example of these efforts, though Chesnutt comes to different conclusions than either Washington or Du Bois.

Chesnutt's Conjure Stories carefully balance an almost minstrel-like essentialism of some of the main characters with an underlying celebration of African American survivance.[19] In "Superstitions and Folk-lore of the South," Chesnutt writes that Joel Chandler Harris presents in his "Uncle Remus" stories "plantation stories which dealt with animal lore" as opposed to those that engage with "so-called conjuration."[20] Chesnutt describes conjure as a combination of African traditional knowledge, European ghost stories, and Native American legends. "To many old people in the South, however, any unusual ache or pain is quite as likely to have been caused by some external evil influence as by natural causes."[21] A conjurer is, according to Jeffrey Anderson, "a professional magic practitioner who typically receives payment in return for his or her goods and services."[22] Chesnutt also places conjure on a spectrum of belief between traditional religions and superstitions. The element of superstition allows conjure to be dismissed as another opportunity to diminish the African American community from mainstream success. Chesnutt, on the other hand, works to credit achievements, as is visible in his descriptions of conjure as well as animal husbandry. Julius, for example, manages to acquire something tangible at the conclusion of each of his tales. He manipulates his audience's sentimentality for his own benefit, and readers applaud him for that success. Chesnutt understands regionalism in practice through his use of dialect and conjure; this regionalism appears in his discussion of southern agrarianism as well. The characters in Julius's tales within the *Conjure Stories* are also able to find a measure of success—or the ability to survive within the slavery system's form of southern agriculture—by using their wits and knowledge of conjure as part of a larger cultural tradition, one that includes southern cuisine, which Chesnutt parodies while also celebrating.[23]

There are several stages of cultural commentary in the *Conjure Stories*. The work of food studies in these readings exists at the intersection of ecocriticism, posthumanism, and critical race studies. I employ these theories against the agricultural setting of the stories, therefore. Indeed, agriculture is the environment, engaging with nature as a fundamental part of its function. Too often, intensive agriculture has abused the environment in ways that few people during the nineteenth century understood, though its results were becoming apparent even then. Chesnutt describes the theme of ecological toxicity—the dangers of intensive agriculture without a balanced concern for human or environmental consequences—within southern agricultural practices as part of his analysis of southern regionalism in what Lawrence Buell describes as the "physical environment humans actually inhabit," a space where humans and the nonhuman world are "biotically imbricated" and "modified (like it or not) by techne" simultaneously.[24] This reading applies especially to agriculture, where nature and "techne" have always been combined. In Chesnutt's case, the environment within North Carolina's agriculture remained particularly toxic following Reconstruction, as economic expansion was by that point moving largely west rather than south, leaving farmers in the South to make do with an impoverished environment.[25] Furthermore, in Chesnutt's stories, humans, plants, animals, and the land have dissolved their distinct boundaries through conjure, and posthumanism provides a lens to engage with this fluidity. The fact that humans and the nonhuman world can easily lose these divisions in Chesnutt's stories shows the intimate connection people have with their food. Chesnutt describes unsustainable agricultural practices during the time of slavery and its aftermath when animals and humans are brutalized for profit.[26]

I add critical race studies to these ecocritical and posthumanist food studies readings, therefore, to challenge the treatment of people of color as well as nonhuman labor and food animals in Chesnutt's stories, building on Michael Omi and Howard Winant's concept of racial formation.[27] The theory of critical race studies works to invert "the terrible historical legacy of making people of color signify the natural, as a prelude to exploiting both," writes Paul Outka, adding race theory to ecocriticism; I agree with Outka that Chesnutt's *Conjure Stories* are a "most powerful and subtle critique" of Euro-American pastoralism.[28] As I focus on "The Conjurer's Revenge" and "A Victim of Heredity," I engage with stories that deal closely with animal husbandry within the nineteenth-century U.S. South's abusive system of agriculture, large-scale and small. Also, these are two of the only Chesnutt tales that mention beef or cattle, though in far less detail than either pigs or chickens, making them a useful complement to the national narrative of agri-expansion and cattle.[29]

Conjuring a Mule as Revenge for a Pig

Chesnutt's "The Conjurer's Revenge," like all of his *Conjure Stories*, speaks directly to the pseudo-anthropological works of the time, such as Harris's "Uncle Remus" stories and Albion W. Tourgée's *Bricks without Straw* (1880).[30] While appearing to model the style of Harris, Chesnutt subverts the plantation genre of the "Uncle Remus" stories, challenging the racism of slavery and Reconstruction in his "Uncle Julius" tales. According to Henry Louis Gates Jr. and Maria Tatar, "No one was more successful at recuperating the Negro folk tradition from 'the plantation tradition' than Charles W. Chesnutt."[31] In "The Conjurer's Revenge," John is deciding between buying a mule or a horse for farm labor. Julius's tale describes a conjure man turning one of the story's main characters, a man named Primus, into a mule as revenge for the theft of a piglet. The mule-as-Primus has several adventures, including avoiding labor when he is sent to carry a portion of beef. Finally, though, the conjure man realizes he is ailing and decides to return Primus to his human form before he dies. He dies before he can complete the transformation, though, and the human-Primus is left with a club foot. At the end of Julius's tale, John has been convinced to buy the horse, though he comes to realize this is the wrong choice after all. Chesnutt brings agricultural commentary into "The Conjurer's Revenge." As cattle had become a dominant image for westward expansion in the United States, the pig came to represent a far more significant element of the agriculture of the antebellum South than the cow.

A review of the history of pigs is important to an understanding of Chesnutt's agricultural work in the story. In the 1800s, Chinese pigs were brought to Europe (and then to the Americas) to improve stock in what was a key step in the extensive industrial production of pigs in particular. Chinese pigs had been domesticated eight thousand years earlier than European ones. These separate lines of production between the two global zones influenced society in both. In China, a pig stood for wealth on a small scale, as seen in the fact that the Chinese character for "home" is the character for "pig" beneath that for "roof."[32] England's pig husbandry developed later, that nation finally having its own agricultural revolution in the eighteenth century. From there, pigs came to the United States. Pork was the easiest—and therefore cheapest and most available—meat source for U.S. Americans to produce for three centuries. To a large extent, this changed with agri-expansion of the cattle industry in the nineteenth century. This agri-expansion, though, would not come to the South in the same way, and beef would remain a relatively rare, and therefore high-status, food.[33]

In Julius's framed tale within "The Conjurer's Revenge," Primus initiates the narrative tension by stealing a shoat—a piglet—from the local conjure man. "De shote des 'peared ter cha'm Primus, en fus' thing you know Primus foun' hisse'f 'way up de road wid de shote on his back" (The shoat just appeared to charm Primus, and first thing you know Primus found himself way up the road with the shoat on his back).[34] This theft creates the primary conflict of the story, as it is his "revenge" that causes the conjure man to turn Primus into a mule and get him sold to another plantation in the first place. There is a relatively simple logic behind the importance of a young pig in this case. If raised tame from a shoat, pigs are a particularly simple animal to grow to an appropriate size for slaughter and production. Omnivores, they can be fed kitchen scraps and other waste feed, as opposed to cattle, which require a grass-based diet as strict herbivores. Pigs grow rapidly, getting to adult size and ready for slaughter in less than a year. Pig meat can be preserved, as well, through salting, smoking, and curing in a far simpler manner than beef. In addition, the conjure man calls himself a "cow-doctor," or veterinarian, providing even more support to his understanding of animal husbandry. Thus, when Primus steals the conjure man's young pig he is taking something known to be of great value, an investment.[35]

The conjure man quickly works "his roots" in order to discover the thief: "One mawnin', a day er so later, en befo' he got de shote eat up, Primus didn' go ter wuk w'en de hawn blow" (One morning a day or so later, and before he got the shoat eaten up, Primus didn't go to work when the horn blew).[36] In a dramatic moment, the conjure man has discovered that Primus was the culprit, and that he had already started eating the shoat. Therefore, Primus is being punished for more than merely theft. In addition, he doesn't properly use his investment, as the conjure man would have done. The revenge is worked before the shoat was "eat up," meaning that Primus was in the process of eating the young piglet, rather than waiting for it to grow into a large pig that could be slaughtered when it had reached full size and contained much more consumable meat. This is another example of Chesnutt's theme of poor economics in terms of agricultural sustainability—in this case the efficient and sustainable raising of a shoat into a year-old fattened pig. The conjure man would know this, and he is therefore revenging more than a few meals' worth of meat, but rather a sizable amount.

This tale about a conjure man avenging the theft of his pig is also one of Chesnutt's few stories that mentions beef. As is the case whenever Chesnutt describes beef in his stories, it is shown to be something remarkable: "Mars Dugal' had kilt a yearlin', en de naber w'ite folks all sont ober fer ter git some fraish beef, en Mars Jim had sont 'Dolphus fer some too" (Master Dugal' had killed a yearling, and the neighbor white folks all sent over for to get some fresh beef, and Master Jim had sent 'Dolphus for some too).[37] The idea of fresh beef, rather than other types of meat, is bringing representatives of the entire Euro-American community to Dugal's plantation. Only these few fortunate neighbors are able to obtain such a privileged foodstuff. It is to be transported by the enslaved laborers, but only that; they are not to be seen consuming this high-status meat. Beef signifies wealth and is largely inaccessible to the enslaved populations described in Julius's tales. Beef was a privileged meat, catapulting its consumer into the ranks of true citizenship; on the other hand, the lack of beef consumption could remove a body from that citizenship. Meanwhile, the consumption of pork, ham, bacon, and other versions of pig meat became significantly associated with the South—implying a kind of liminal citizenship status. Rather than on the national scale, the pig becomes an important regional signifier, like Julius's dialect. Chesnutt laments the loss of inclusion into true nationhood felt by the small-scale southern farmers at the time he was writing, while also celebrating the region through its culture, as is shown in Chesnutt's descriptions of various levels of meat rations in "The Conjurer's Revenge."[38]

If beef is the highest-status food in southern cuisine, as seen in this story, pork is the lowest-status meat, though still important enough for Primus to be turned into a mule for the theft of a shoat. A significant protein, pork would still be preferred over no substantial protein at all. Chesnutt focuses on the rations of the enslaved laborers in several stories, including this one. The meat of pigs, specifically bacon, is given as the normal rations to enslaved laborers throughout Chesnutt's stories. In this tale, another man, Pete, is asked to return the mule to the conjure man so that he can remove the conjure. First, though, Pete eats: "Pete didn' know w'at de cunjuh man wuz dribin' at, but he didn' daster stay way; en so dat night, w'en he'd done eat his bacon en his hoe-cake, en drunk his 'lasses-en-water, he put a bridle on de mule, en rid 'im down ter de cunjuh man's cabin" (Pete didn't know what the conjure man was driving at, but he didn't dare to stay away; and so that night, when he'd eaten his bacon and his hoe-cake, and drunk his molasses and water, he put a bridle on the mule, and rode him down to the conjure man's cabin).[39] These three food items—bacon, cornmeal, and molasses—appear throughout Chesnutt's stories as basic rations for North Carolina laborers. The meat of pigs is omnipresent in southern culture, as opposed to the privileged beef that 'Dolphus must carry for his Euro-American neighbors.[40]

The similarity of Primus's mule-life compared to his human-life is made clear in the beef scene that focuses on privilege and meat production: "'Dolphus lef' de mule stan'in' in de ya'd, en went inter de smoke-house fer ter git de beef. Bimeby, w'en he come out, he seed de mule a-stagg'rin' 'bout de ya'd' ('Dolphus left the mule standing in the yard, and went into the smoke-house for to get the beef. Bimeby, when he came out, he sees the mule a-staggering about the yard).[41] The mule, with its human intelligence, had opened a barrel and become drunk on scuppernong wine. "'Dolphus had ter take de beef home on his back, en leabe de mule dere, 'tel he slep' off 'is spree" ('Dolphus had to take the beef home on his back, and leave the mule there, until he slept off his spree).[42] The mule, unlike the human Primus, has been able to consume wine to excess. Another human, 'Dolphus, is working harder because Primus-as-mule is working less. 'Dolphus carries the beef on his back, becoming a parody of a mule, an iconic beast of burden—and also a representative animal of the southern region—in much the way oxen came to represent the plowing of the Great Plains. The bitter humor of this tale is in the treatment of a human being as similar to—if not worse than—the treatment of an animal.

In addition to his description of the value of beef and pig as meat sources in the South during this time, then, Chesnutt also manipulates the human-animal binary through the use of a mule. Instead of the oxen helping to plow the West, Chesnutt introduces mules and horses as primary laboring animals in his *Conjure Stories*. In this story, Primus is turned into a mule, and not, for example, an ox. Mules are often used in stories in the African American South. Zora Neale Hurston writes a variation of Chesnutt's story of men and mules, in which a mule tells a man to "Wanh-uh! Go preach! Go preach! Go preach!"[43] The man in Hurston's story is then unsuccessful in his attempts to become a preacher, eventually becoming a plowman. Mules have a European history in folktales as well. Mules are more than symbols of the South for conjure and folktales; they are also representative in terms of their real labor. In the quote from the beginning of this chapter from "Mars Jeems's Nightmare," Solomon places horses before mules in his agricultural rankings, as he places cows before pigs. However, the mule was also the animal of choice for much of the manual labor involved in southern agriculture. Robert B. Stepto and Jennifer Rae Greeson note the racism behind the symbol of the mule in the agricultural South: "An inter-species cross between horse and donkey, mules are designed by their breeders to be infertile, tractable, capable of hard labor, and able to withstand abuse."[44] In "The Conjurer's Revenge," the racism behind treating humans like mules is emphasized to dramatic effect.

By the end of Julius's tale, both Primus and the conjure man are worse off. The conjurer has died, having been poisoned accidentally while trying to make amends for turning Primus into a mule. Meanwhile, Primus is transformed back into a human, but because the conjurer died before he could return Primus fully back into his human form, he is left with a club foot. Stepto concludes that Primus's imperfect return to human form represents "one of Chesnutt's most serious themes: the great improbability of a slave's escape from slavery's 'crippling' effects."[45] No one is blameless in this story, but the characters are also punished beyond their desert. John is committed to making a profit from his vineyard and other crops, notably focused on a northern trade; however, by the end of the story he has shown some character development. John has wanted a mule, while Julius recommends a horse. After the conclusion of Julius's conjure tale, John, perhaps subconsciously, follows Julius's recommendation, buying a horse rather than a mule for labor. He does this even though he says of himself, "I was not a very good judge of horse-flesh," and even though the horse ends up being a poor investment.[46] In this way, then, John is able to change somewhat from the non-actor of the previous stories to the one who acts on Julius's recommendation (a role previously played by Annie in the stories). John admits that he has learned from this experience. Stepto and Greeson observe that the various versions of this story—from its 1889 publication in *Overland Monthly* to its 1899 form in *The Conjure Woman*—change the character of Annie primarily, Chesnutt's moral compass figure. Her reactions vary between the editions, showing the difficulty of finding an easy "reading" of a story about the dehumanization of enslaved laborers through men, pigs, and conjure as the nation's agricultural systems intensified following the Civil War, Reconstruction, and Manifest Destiny.[47]

"The Conjurer's Revenge" has been noted as being different from the other *Conjure Stories*, as it is almost entirely about African Americans engaging with each other. Euro-American characters are not the focus of this story. Chesnutt uses conjure to describe a particular hybrid of a mule and human, as well as its personality. He focuses on the revenge taken because of the theft and misuse of a pig—a representative southern livestock animal—as well as the relative scarcity of beef for any but the most privileged in the region. "The Conjurer's Revenge" is one of Chesnutt's only stories, indeed, to engage with both cattle and pigs in terms of their representative values. Cattle, the representative U.S. livestock, is practically not an option for Chesnutt's characters, Euro-American or African American. "The Conjurer's Revenge" then shows the complicated agriculture of the post–Civil War South, a theme that will continue in Chesnutt's yet more complicated story, "A Victim of Heredity."[48]

The Bull Market of Chickens

In a later conjure tale, "A Victim of Heredity: Or, Why the Darkey Loves Chicken," published in 1900 in *Self-Culture Magazine*, having been written for but not included in the 1899 *The Conjure Woman* collection, Chesnutt comments again on race in terms of the economic forces of intensive animal husbandry.[49] The story begins with John catching a chicken thief, Sam Jones, Julius's nephew, and detaining the thief in the smokehouse until law enforcement can arrive and take him away to the penitentiary. John then asks Julius why his "people" love chicken to such an extreme. After a discussion in which Annie accuses both John and Julius of overgeneralizing the ways of an entire community, Julius tells an origin story about the regional love of chicken. In his tale, pigs, cattle, and chickens are brought into a market-style economy—beyond the more sustainable barter system—as Peggy, a conjure woman, and Tom, the nephew of the plantation owner, help to ruin the story's villain character, Donal' McDonald, through a perfectly executed corner on chickens. This is a striking commentary on agriculture in the land of cheap and plentiful meat. In this case, a change that causes too drastic a shift toward commodity-based animal husbandry results in failure and collapse. Often in Chesnutt's stories the African American characters suffer after conjure and other unnatural events; in "A Victim of Heredity," though, it is the Euro-American villain who is ultimately brought to his comeuppance through conjure and his own hubris. Indeed, the lives of the African American characters show marked improvement in their diet following the events of this tale, though they remain solidly within the bounds of slavery.[50]

Chickens are a differently symbolic animal than either cattle or pigs in terms of cultural representation. Literary works provide ideal ways to engage with African American culture, including African American food culture, described in the work of Psyche Williams-Forson.[51] She introduces the history of the chicken (*Gallus domesticus*) as domesticated in 1400 BCE, coming to the Americas from Asia, not unlike the pig, though in this case most likely from Java and India. These chickens became part of the surplus wealth of colonists in the Americas, and when enslaved people began arriving from West Africa, chickens would come to make up a significant part of their agricultural and culinary experience, as well. Chicken became, according to Williams-Forson, "commonly affiliated with black expressive culture."[52] Like pigs, chickens were far easier to produce than cattle. Small-scale farmers could maintain a chicken coop with relatively little effort, providing their families with protein in the form of eggs as well as whole chickens. However, as beef became more readily available to U.S. consumers, chicken became far less preferred, sometimes below even pigs as an acceptable form of meat. As chicken was being rejected more often by Euro-Americans, and while it remained a useful protein source, it grew to some extent in importance for a southern, and largely African American, culinary culture.[53]

In this story, John establishes a hierarchy of food quality and appropriate punishments in the first few paragraphs—specifically regarding African Americans. "I suffered more or less, from time to time, from petty thievery," he states stoically. After noting that he is used to forgiving the theft of all manner of vegetable foods, John concludes: "One summer, after several raids upon my hen-house, I determined to protect my property. I therefore kept close watch one night, and caught a chicken-thief in the very act. I locked him up in a strongly-built smokehouse, where I thought he would be safe until morning."[54] Vegetable crops are valued less than animal foods grown on-site, even chickens. For the theft of meat, therefore, there is less forgiveness—rather, there is a long detention first in an inhospitable smokehouse (which were never intended for humans), and then in a penitentiary. The smokehouse itself, as noted by Stepto and Greeson, among others, connects John's actions directly to characters in stories like "Po' Sandy" and "Dave's Neckliss," who used their smokehouses to torture other humans.[55] The question of "heredity and environment" is hinted at in this instance—with John clearly coming down on the side of heredity. He even wonders to Julius if the lust for chicken is "in the blood" for African Americans. John had spoken in the previous paragraph of "the larger opportunities of freedom" that would allow emancipated men and women to "improve gradually and learn in due time to appreciate the responsibilities of citizenship."[56] After this general statement, though, John soon reverts to punitive measures; his goodwill stretches thin quickly. Using Wallach's analysis of beef as requisite for elevated status discussed earlier in terms of the Tuskegee Institute, John's reactions regarding the possible theft of a mere chicken shows the difficulties African Americans in the South encounter seeking even a more basic form of meat-protein enfranchisement. In *Up from Slavery* (1901), Washington describes his mother's theft of a chicken to feed her children: "Some people may call this theft. If such a thing were to happen now, I should condemn it as theft myself. But taking place at the time it did, and for the reason that it did, no one could ever make me believe that my mother was guilty of thieving."[57] By the end of Chesnutt's story, as well, the accused is set free, declared not guilty of theft by Annie, at least, as she has absorbed Julius's tale and releases Sam before law enforcement can arrive.

Julius's tale in "A Victim of Heredity" involves the conjure woman, Peggy, an emancipated woman who wishes to give Tom McDonald a conjure favor after he saves her from drowning. Her opportunity arrives when Tom's uncle, the lazy Donal' McDonald, a former overseer who has worked his way up to becoming a plantation owner, comes to Peggy for a conjure charm to make him able to feed his enslaved workers for less money as a continuation of his extreme frugality and overall disregard for human life, especially African American human life. (He even tries to steal back the coin he uses to pay Peggy with, but she has prepared for this eventuality, conjuring it to be too hot for him to touch.) Peggy makes him a potion to put into his laborers' rations that will allow him to get away with feeding them exactly half their normal rations for one week, without any seeming adverse effects. This move on Donal's part is a kind of agricultural innovation. He is experimenting with ways to increase production by reducing expenditures, an important move in U.S. agriculture as it expanded and commodified throughout the nineteenth century and beyond. In the example in this story, the timeline is antebellum; clearly, however, some version of this impulse would have been encouraged in actual practice beyond the time that Julius describes. John himself figures out the costs and benefits as he decides to take over the McAdoo plantation in North Carolina in the collection's first story, "The Goophered Grapevine." Chesnutt describes Donal' as a villain in "A Victim of Heredity," but it is significant that his agricultural efforts largely would have been encouraged by society.

After Donal's visit, Peggy tells Tom to gather "all de money you kin rake en scrape, en you git all de credit you kin; en I ain' be'n conj'in' all dese yeahs fer nuffin, en I'll len' you some money. But you do des ez I tell you, and doan git skeert, en ev'y-thing'll tu'n out des exac'ly ez I say" (all the money you can rake and scrape, and you get all the credit you can; and I haven't been conjuring all these years for nothing, and I'll lend you some money. But you do just as I tell you, and don't get scared, and everything will turn out just exactly as I say).[58] If he trusts her, she tells him,

Tom will get all his money back, plus money from Donal', and he will finally be able to marry the woman he loves—of whom Donal' disapproves. Peggy's conjure potion works, with such remarkable results that Donal' decides to carry the conjure even further, halving his workers' rations multiple times instead of just once. Eventually, this stops being effective in the way Peggy has described; the laborers grow weak and unable to work—they in fact seem to be coming close to starving to death. Donal' then returns to Peggy desperate for a way to resolve this new problem through conjure. To his entreaties, she replies that the problem is of his own making. Her temporary solution for him, then, is to feed his enslaved workers particularly well. Instead of the meager rations of bacon, cornmeal, and molasses, she suggests better cuts of pig, then roast beef, and finally chicken. At this point, Julius returns to Peggy and Tom's financial scheme. They had cornered the market on chickens. Between the two of them, they make Donal' pay almost all his wealth in order to obtain enough chickens to feed his starving enslaved workers. This scheme ultimately ruins Donal', and Tom takes over the land.

The weekly rations for people on Donal's plantation amounts to "a poun' er bacon en a peck er meal en a qua't er merlasses" (a pound of bacon and a peck of meal and a quart of molasses) each.[59] The same diet is noted for Pete in "The Conjurer's Revenge." Chesnutt describes the common rations of enslaved people during the antebellum period as being largely composed of bacon, cornmeal, and molasses in both stories.[60] This is the baseline of food required for a person to stay healthy enough to do physical labor. Peggy's revenge scheme improves the diet of the workers, providing them with greater quantities of more expensive foods: "roas' po'k," then "roas' beef," and, finally, "chick'n."[61] The pig was already part of normal rations, but quickly Peggy's recommendations move beyond a lower category of meat to beef, a higher-quality and more limited agricultural commodity, and finally to chicken and her successful corner. In the story, beef stands out against the standard rations of bacon and other pig products, or something small like chickens, which are relatively cheap and therefore simple enough to be acquired in bulk. There simply aren't that many cows in the region. Roast beef, therefore, becomes the food requirement in Peggy's plan that lasts the shortest amount of time, and its scarcity causes it to be more expensive, helping to bankrupt Donal' without the necessity of her investing a great deal of her own and Tom's money. By the end of the tale, it has been established that the South had a relative dearth of cattle and plentiful pigs and, especially, chickens.

Peggy is working a corner of the market in chickens that would come back to hurt Donal', as if she knows all that will transpire, that Donal' will abuse his "goopher mixtry" and that the laborers would be brought to near-starvation levels. This might seem reprehensible behavior on Peggy's part; however, it might also be the case that her goopher functioned in such a way that the workers seem to be weak and starving, while in fact they feel at least a little better than they appeared. If we follow the logic of conjure, this is possible. Peggy shows her strength and intelligence as she manipulates Donal's cheapness—knowing he would take advantage of the goopher and abuse it. The chicken scheme wouldn't have worked, after all, if he only feeds his enslaved workers the half rations for the one week. However, because he abuses the goopher, reducing the rations of his workers for multiple weeks, he gets to the point where the enslaved workers are too weak to do much—and that is what helps Peggy move on to the pork-beef-chicken part of her economic plan. Significantly, Peggy succeeds in a market-based world, not a barter economy, as is more customary in the *Conjure Stories*. Peggy's final recommendation is for Tom to continue to feed his enslaved workers chicken at least once a week, referencing the lasting damage Donal' has done with his manipulation of her conjure. In this way, she succeeds in improving the diet of the local African American community in a lasting way. After ruining Donal' and helping Tom, Peggy is also rewarded with "all de chick'n en wheat-bread she wanter eat, en all de terbacker she wanter smoke ez long ez she mought stay in dis worl' er sin en sorrer" (all the chicken and wheat bread she wants to eat, and all the tobacco she wants to smoke as long as she might stay in this world of sin and sorrow).[62] Ultimately, Peggy herself is treated to plentiful chicken dinners, a ration above the basic.

The story ends with Annie thinking about the issue of "heredity and environment," although, she admits, "I have not been able to get everything reasoned out."[63] At the same time, John's changing attitude regarding how to punish the chicken thief shows some character development on his part, while Annie's reaction to the story is seen as another form of patronization.[64] "A Victim of Heredity: Or, Why the Darkey Loves Chicken" is often read as too closely paralleling minstrel stereotypes of African Americans in the nineteenth century. Eric Sundquist challenges this criticism, though, noting that the story "asserts not just that environment, rather than 'blood,' is the cause of behavior, but also that the cultural origins of such stereotypes in racism can be rooted out and overturned," in this case through the actions of the conjure woman, Peggy, as much as Julius's influence over John and Annie.[65] This, along with the failed attempt at an inherently flawed kind of innovative market-based agriculture by Donal', add strength to his reading. Nature, or "blood," is not seen in this story at all; rather, it is the machinations and strategizing, the "techne" of agriculture. Also, as cattle had become the representative animal of U.S. agriculture by this period, much of the South had become economic "victims"—making small-scale agricultural victories like Peggy's and Julius's more significant when they occur.[66]

Julius explains the reason why "his kind love chicken" by "contrasting the minor theft to the serious crime that the master commits in trying to cheat his slaves of one of the most basic forms of recompense they could receive for their labor," according to Glenda Carpio: food.[67] "A Victim of Heredity," like some of Chesnutt's other essentializing stories, does contain a minstrel-like tone. Sundquist calls this one of Chesnutt's "most bitter tales" as it has "a signifying but bittersweet critique of the stereotypes' pervasive presence in American life and the consequent difficulty of the black artist's escaping their denigrating influence."[68] In this story, the way Chesnutt deals with agriculture and diet in particular makes an interesting commentary on the cultural stereotype. Chesnutt does equate African Americans with chickens in a way that draws dramatic attention to a racist trend in U.S. culture. Meanwhile, the agricultural lineage of Chesnutt's style is seen in his contrast to Harris, who longs for an antebellum agrarianism in his idealized plantation slavery. Chesnutt, on the other hand, had no such fantasy. He knows that the agricultural system was broken in the South, a region that happened to be lacking large-scale cattle production at a time when cattle were becoming particularly exploited in the nation, and he shows this throughout his stories—specifically through the use of broken forms of animal husbandry and a corrective form of conjuration. In this story, food animals are part of a commentary on unsustainable treatment of humans, animals, and the land in large-scale plantation agriculture based on coerced labor. An already unsustainable system is pushed past its breaking point and fails completely—in this case to the benefit of the conjure woman and the laborers themselves.[69]

Conclusion: Goophering Southern Agriculture

Charles Chesnutt's *Conjure Stories* describe food and agriculture throughout: readers see people grow food, prepare and consume food, use food as currency, and become transformed into food items through conjure and human cruelty. Pigs and chickens are the most commonly described food animals in these stories. In addition, though, grapes are discussed, as are watermelons, persimmons, okra, and sweet potatoes, and non-food consumables such as sugar in its various forms, tobacco, turpentine, and mineral water. At the same time, the oxen predominant in narratives of westward expansion are replaced in these stories by horses and mules. Julius is a driver, so he engages the horses frequently in both past and present timelines, and "The Conjurer's Revenge" focuses on mules in particular.[70] Cattle and oxen require grazing lands that are not practicable in the South. In terms of prepared food, beef is mentioned in certain stories, but never to the degree of either pig or chicken. When cattle do appear, they are symbols of wealth and status. Susan Fenimore Cooper discusses her concerns with deforestation for grazing as a form of faulty resource management along the frontier, and this concern appears as well in Chesnutt, as characters like the experimental agrarian John focus on crops like his grapevines rather than grazing livestock in his recovery farm work.[71]

Moreover, I argue, the issue of appetite—and what is being hungered for, the manner in which it is grown—is significant in these stories in terms of dealing with questions of value. In consumption and transformation, there is a divide between nurture and nature, environment and biology. For Chesnutt, the cattle, pig, and chicken divide is another way to examine his themes of race, nature, and agriculture in the post–Civil War South. The contemporary vegan cook Bryant Terry argues that African American communities in the South should be valued to a greater extent than they are for their culinary contributions from Chesnutt's time to the present: "There is a notable failure to (1) acknowledge that the modern world is indebted to ancient Africans for basic farming techniques and agricultural production methods; (2) appreciate the agricultural expertise (rice production), cooking techniques (roasting, deep-frying, steaming in leaves), and ingredients (black-eyed peas, okra, sesame, watermelon) that Africans contributed to new world cuisine; and (3) recognize the centrality of African-diasporic people in helping define the tastes, ingredients, and classic dishes of the original modern global fusion cuisine—Southern food."[72] While Terry encourages a move to a more plant-based southern cuisine, Chesnutt's stories are largely associated with the meat of pigs and chickens, as opposed to beef. Otherwise, Chesnutt seems to agree with the cultural reclamation work in Terry's argument. Chesnutt shows the complicated economy of southern agriculture—specifically related to a history of "making do."[73] While he does this, he also shows the lived experience of people and elements of their food culture in this region.

By Chesnutt's time, agri-expansion had decidedly moved west for cattle, the South being the region chosen for intensive production of a variety of other cash crops, as well as the domestic pigs and chickens. Chesnutt's *Conjure Stories* use animals, agriculture, and other natural elements—more than any of his other works—to present his challenges to racism and the ongoing violence against African Americans at the end of the nineteenth century. *The Conjure Woman* remained Chesnutt's most successful publication during his life, though he would actively continue writing for years following its release (see fig. 16). He followed *The Conjure Woman* closely with a biography, *Frederick Douglass* (1899), the collection *The Wife of His Youth and Other Stories of the Color Line* (1899), and the novels *The House Behind the Cedars* (1900) and *Marrow of Tradition* (1901). By this later work, he was more directly challenging racial conditions in the country, and he grew less popular commercially. He published little after that, devoting himself largely to activism as a member of the National Association for the Advancement of Colored People. Race and agriculture were foundational concerns for Chesnutt throughout his *Conjure Stories*, his later writings, and his life in activism.

Fig. 16. Charles W. Chesnutt at age forty, ca. 1897–98, shortly before the publication of *The Conjure Woman*. Cleveland Public Library, Fine Arts and Special Collections Department, gift of Helen Chesnutt.

Chesnutt draws upon this marrow of cultural iconography of cattle, and the lack thereof throughout his writings, to show the struggles and creative solutions of his southern characters, including in their cuisine and agriculture. I argue that though Chesnutt does not use the cow as his symbolic animal in these stories, the absence of cattle is just as striking in ways that add nuance to the narratives involved in *Cattle Country*. Chesnutt shows that race and the environment were interconnected during the nineteenth century through the unsustainable system of intensive agriculture that is too large-scale for humans—especially African Americans and other marginalized peoples—and for nonhuman animals to survive well, but also not large-scale enough to compete economically with the U.S. agricultural system expanding west. The South of Chesnutt's *Conjure Stories* was being excluded from the ecological and economic changes occurring as agriexpansion moved farther west over the nineteenth century. Moreover, Chesnutt uses his chosen food animals to show injustices against race and nature following the Civil War and the failure of Reconstruction in ways that complement the arguments of María Amparo Ruiz de Burton. The symbolism of cattle continues into the early twentieth century both regionally and globally. In the following chapter, Upton Sinclair makes an impassioned argument against intensive U.S. agriculture as it was at the turn of the century, specifically regarding cattle and other forms of livestock, and Winnifred Eaton examines cattle as representative animals on the global stage.

SEVEN

Industrial-Global Cattle in Upton Sinclair and Winnifred Eaton

Upton Sinclair demonstrates the culmination of the nation's nineteenth-century cattle industry's expansion and intensification in his muckraking classic, *The Jungle* (1906). The human-made elements of production are emphasized in the passage of millions of animals: "There were groups of cattle being driven to the chutes, which were roadways about fifteen feet wide, raised high above the pens. In these chutes the stream of animals was continuous; it was quite uncanny to watch them, pressing on to their fate, all unsuspicious—a very river of death."[1] The chutes are described as "roadways," elevated structures "raised high," impressive feats of engineering. However, the natural world is represented as well; these livestock-filled chutes are also a "tide," a "stream," and a "river" (see fig. 17). This bloody abundance is destined to feed the world, specifically the U.S. American world. One of these Euro-Americans is represented by Tojin, the male protagonist of Winnifred Eaton's Japonisme novel *Tama* (1910). Eaton describes Tojin's dietary requirements early in the novel: "Food sufficient for six ordinary mortals must be prepared for his individual consumption. Raw meat and game, lightly scorched before fire, were essential."[2] Only the production of meat at the scale described in Sinclair could, it seems, satisfy the demands of

Tojin. In Eaton, cuisine becomes an essentializing feature of both Japanese and U.S. cultures, a site of global imperialism. Tojin is contrasted to the "ordinary mortals," the Japanese villagers, who are able to live on far less food, specifically on far less meat. In addition to cuisine, Eaton describes the natural world in this case in the manner in which the meat is prepared. Tojin consumes hunted "game," "lightly scorched" on an outdoor fire. A little later in the novel, as Tojin is being escorted to his new home, part of the general fanfare involved in his journey involves his general dietary needs, specifically for quantities of food: "Every little town and hamlet sent to him on its outskirts deputations of high officials. There had been feasts here and banquets there."[3] In these scenes, Eaton emphasizes the focus on red meat that may or may not have included cattle that a Euro-American man is presumed to require. The scale is entirely different in the two works discussed in this paragraph, but the emphasis on the quantities of red meat required to sustain U.S. consumers is comparable at the local, national, and global setting.

Overall, *Cattle Country* has interrogated some of agri-expansion's direct effects on humans and the nonhuman world throughout the literature of the nineteenth century. In this chapter I pursue that argument into the early twentieth century, as authors continued to challenge agri-expansion's influence on people, animals, and the land on the industrial and global scale. I engage with two seemingly disparate authors—Sinclair and Eaton do not seem to have known each other, nor do they have similar biographies—who have cattle production in common. Indeed, even the different ways in which they use cattle and other domestic animals is significant: Sinclair wrote one of the muckraking classics of U.S. history in his survey of the cruelty and misuse of resources—animal and human—by the Beef Trust in Chicago, as an example of the packinghouses of the major midwestern cities of the United States; Eaton, on the other hand,

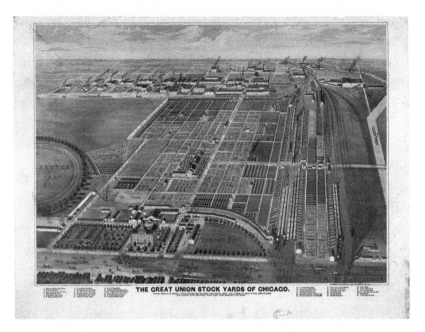

Fig. 17. *The Great Union Stock Yards of Chicago*, 1878. This
Charles Rascher image of Chicago's Union stockyards gives
a sense of the magnitude of the cattle-processing operation.
Library of Congress, Prints & Photographs Division.

wrote novels set in Japan, the United States, and Canada, in which
food plays an important role, specifically beef and other forms of
meat. Therefore, I study Sinclair's *The Jungle* as well as Eaton's
Tama and *Cattle* (1924), as they confront industrial agriculture
and globalization, respectively.

The work of food studies applies both to Sinclair's writing on
the Chicago meatpacking industry at the turn of the twentieth
century and Eaton's romanticized novels set in Japan and Canadian cattle ranches as they deal with issues of industrialization
and early agricultural globalization. Both authors took advantage of the literary opportunities of their time to use narrative to
engage with the production of meat in the early twentieth century. Throughout the nineteenth century and into the twentieth,

the United States underwent drastic shifts in demographics as urbanization, immigration, and continued westward migration contributed to the rise of literary regionalism—literature that focused on detailed descriptions of key elements of regional culture, including cuisine. In addition, food production made a sizable and lasting environmental impact with a shift to cities like Chicago. Agriculture had an impact not only on the environment but also on other global issues, such as economics. Technological advances in the nineteenth century like the telegraph had a dramatic effect on agriculture, as well. Chicago was ideally located for both its economic growth through the industrial meat production and the exposé written about the abuses connected to that boom. Moreover, the global elements of both Eaton's and Sinclair's work engage with xenophobia and cosmopolitanism. I apply the concept *cosmopolitan* as it is developed by Kwame Anthony Appiah: an expression of mutual respect across ethnic, political, or national borders. The Lithuanian, Polish, and African American characters in *The Jungle* are made comparable through their oppression by the Beef Trust even as they are driven to fight against each other for meager opportunities. Meanwhile, the Japanese and Euro-American characters are essentialized as fundamentally separate in *Tama*, while Canadian and U.S. ranchers share similar characteristics, as do the English and Scottish, in *Cattle*, characteristics that are largely differentiated from the Chinese cook's character in the novel.[4]

The biographies of Sinclair and Eaton make them particularly appropriate commentators on specific elements of early twentieth-century agri-expansion. Sinclair was drawn to Chicago's story as a Midwest meat-processing boom city, journeying from a small town in New Jersey to visit Chicago stockyards when he was in his twenties in order "to write the *Uncle Tom's Cabin* of the Labor Movement!"[5] Of course, this research led to a phenomenon upon the writing and publication of *The Jungle*.

Following this experience, Sinclair continued to be active polit- ically and narratively for the rest of his life, writing treatises on the oil industry, prison communities, diet, and other controversial topics of his time. He ultimately wrote over eighty novels, essays, and plays, as well as running for governor of California under his EPIC (End Poverty in California) campaign, at which time he was also the victim of one of the nation's first dirty political cam- paigns. Throughout, he was an activist first and an artist second. This combination, though, was effective, especially in regards to industrial meat production. Sinclair famously joked of his success with *The Jungle*, "I aimed at the public's heart, and by accident I hit it in the stomach."[6] Jacqueline Tavernier-Courbin, though, argues, that the public's "misreading" was "not really accidental" but was rather "caused by the vividness of his descriptions of the stomach-turning conditions of work in the Chicago meat-packing plants."[7] Sinclair intended, she argues, to hit the public in both the heart and the stomach. Most importantly, he desired action; he wanted to make people change the way they engaged with the U.S. agricultural industry.

Eaton had a very different biography, one that also allowed her to bring cosmopolitan elements to her writings on agriculture and cattle. A Canadian-born, biracially Chinese-British woman trying to earn a living writing in the United States during a time of severe anti-Chinese xenophobia, Eaton created a presumed Jap- anese identity, focusing her energy on writing about that nation's culture and cuisine. During this time, the United States was in a period of Japonisme, where Japan was being romanticized in works like John Luther Long's "Madame Butterfly" (1898), Lafcadio Hearn's *Glimpses of Unfamiliar Japan* (1894), and Gil- bert and Sullivan's comic opera *The Mikado* (1885).[8] Eaton took advantage of this trend to write novels set in Japan, promoting herself as biracially Japanese and British and therefore as a cul- tural ambassador to Japan, though she had never actually visited the nation. She even gave herself a Japanese-sounding pen name, Onoto Watanna, a name ironically not based on anything truly Japanese, but rather a kind of pen.[9] Amy Ling notes that no less an arbiter of national taste than William Dean Howells enjoyed Eaton's *A Japanese Nightingale* (1902); Dominika Ferens adds that, "given Howells's insistence on writing from experience, his wholehearted endorsement of Onoto Watanna is nicely ironic." What Howells responds to in Eaton's work, according to Ferens, is her liberal use of "not-yet-hackneyed" cultural icons.[10] Her representations of food in her earlier works set in Japan do betray themes of Orientalism and cultural essentialism. After a startling narrative move in which she writes an Irish dialect novel with a cook as her heroine, Eaton returns to her Japonisme subject matter with *Tama* (1910), one of her best-selling novels.[11]

In *Tama*, food and culture are ethnically and nationally, more than culturally, determinate. The novel is foremost a love story involving the implausibly blond, blind, and outcast heroine, Tama, daughter of a Japanese mother and U.S. father, and a U.S. American man, called Tojin in the novel. The lovers have one of their only disagreements in the novel on the issue of meat consumption. At one point, Tojin proposes they eat pigeons living at the nearby temple, proving his need for meat beyond what has been provided by Tama. She categorically refuses, her expression becoming "quite tragic and piteous" as she explains to him, "The gods love them, and I—I may not eat the forbidden meat."[12] Tama's ability to live without red meat seems to be part of her Japanese identity. However, Tojin requires more meat, as part of his U.S. identity. The novel ultimately results in Tama's returning to the United States with Tojin. It is determined that the diet of her Euro-American father and lover would improve her health, too. In other words, Tama joins the Euro-American Tojin in the land of plentiful meat. There is an assumption here that as the Japanese villagers have ill-treated Tama, they have also ill-fed her, something that would be remedied by relocating to the United States.[13] Eaton's *Tama* reinforces the belief that the United States truly is the land of cheap and plentiful meat, a decided benefit for immigrants from the meat-starved regions of the world, such as the biracial main character of *Tama* as well as the European immigrants in *The Jungle*. Following *Tama*, Eaton would develop a more domestic voice in her work set in Canada and would eventually come to include more of her Chinese ethnicity in her writings.[14]

As a multinational and biracial woman who often lived in the cosmopolitan cities of Montreal, Chicago, and New York, as well as a cattle ranch in Alberta, Eaton is particularly well suited to show key elements of cattle production and is worth being analyzed alongside Sinclair. He was writing about the industrialization of the livestock portion of U.S. agriculture at roughly the same time that Eaton was writing of the union of the East and the West through rice and meat. Even as notable a figure as Winston Churchill reviewed Sinclair's novel, commenting that his goal in doing so was "to make it [Chicago's packinghouse conditions] better known."[15] Churchill continues: "Let me say at once that people have no right to hold their noses and shut their eyes. If these things are true, all honour to him who has the power and skill to fasten world-wide attention upon them."[16] Sinclair's ode to the humble immigrant worker is also a shocking exposé of meat-production practices, and the majority of his novel is devoted to describing the manner in which U.S. industrial agriculture is destroying and consuming the workers—primarily recent immigrants—within the meatpacking industry.

Industrial Cattle

Upton Sinclair's *The Jungle*—perhaps the most influential work of U.S. fiction in terms of food reform—forced changes in U.S. legislation of the food industry, after even President Theodore

Roosevelt was disturbed by what was in his breakfast meat. *The Jungle* lies between muckraking journalism and literary natural-ism in its focus on working immigrants and other working-class characters being crushed against the seemingly immovable force of the Beef Trust in the packinghouse district of Chicago around the turn of the twentieth century. However, Sinclair's muckraking account of the great "aggregation of labor and capital" managed to bring about the Pure Food and Drug and the Federal Meat Inspection Acts of 1906, a minor victory on behalf of some of the immigrant populations abused within the system created by agri-expansion that led to the Beef Trust.[17]

In *The Jungle*, Jurgis Rudkus and his family are new to the country, finding work in the meatpacking plants of Chicago as new Lithuanian immigrants. They are slowly crushed—figuratively and almost literally—by the Beef Trust, earning less and owing more and more through a series of accidents and poor luck throughout the course of the novel. Family members are injured and die of preventable diseases. Rudkus himself goes from working a relatively good job in the slaughterhouse, steadily downward through a series of mishaps, until he is forced to work in the dreaded fertilizer plant. Rudkus also becomes a strikebreaking scab, foreman, and hobo at various points in the novel. Similarly unfortunate events befall the other members of his close circle, especially Ona, his wife, cousin Marija, and the maternal Teta Elzbieta. Through it all, Sinclair provides graphic scenes of the production of meat, from the slaughter and dress-ing of cows and pigs to the processing of various meat products. These scenes are Sinclair's most famous, and they are the real reason why the novel had such an impact. Sinclair's discussion of immoral business practices in the most intimate industry of food production was based almost entirely in fact. In particular, his focus on Lithuanian immigrants, that specific ethnicity, was rooted in history. From 5,000 Russians (including Lithuanians) immigrating to the United States in 1880, the number would grow to 250,000 around the time of *The Jungle*'s publication.[18] Upon arriving in Chicago, these Lithuanians and other Russians joined other large immigrant communities like the Italians and Hungarians, often creating unhelpful competition for all groups of workers. In addition, in an alternative agricultural history to that described by Charles Chesnutt in his *Conjure Stories*, the Great Migration, the time when African Americans left the South in large numbers following emancipation, resulted in large numbers of men and women from the South looking for work in Chicago's meatpacking plants as well.[19]

While Sinclair's concerns are almost entirely human, cattle and other animals are also shown to be crushed under the Beef Trust. In particular, cattle are dominant in the novel, almost more of a presence than the humans hired to slaughter them:

> There is over a square mile of space in the yards, and more than half of it is occupied by cattle-pens; north and south as far as the eye can reach there stretches a sea of pens. And they were all filled—so many cattle no one had ever dreamed existed in the world. Red cattle, black, white, and yellow cattle; old cattle and young cattle; great bellowing bulls and little calves not an hour born; meek-eyed milch cows and fierce, long-horned Texas steers. The sound of them here was as of all the barnyards of the universe; and as for counting them—it would have taken all day simply to count the pens.[20]

Sinclair gives the cattle human characteristics here. They are "meek-eyed" and "bellowing," as much a cross section of life as certain neighborhoods in U.S. cities would be in terms of human inhabitants. This diversity and commonality are appealing to the shared sentience within a biocommunity including cattle and people, but it also foreshadows the future of the humans within this system. One form of cattle almost directly led to the decimation of the other in the middle of the North American continent. Additionally, as the cattle are human-like, almost philosophical, in their futile protests, so the people become dehumanized through the agricultural industry that destroys them nearly as systematically as it does the livestock.[21]

Early in the novel, the Rudkuses take a tour of the slaughterhouses, watching in awe as the "wonderful machine" kills cattle, hogs, and sheep with an inhuman efficiency that seems "characteristic of an enterprising country like America," as Sinclair notes.[22] In another scene, the Rudkus matriarch, Elzbieta, works in a sausage-making factory, twisting the casings of the sausage as they come out of the machine. The women doing this work are "so fast that the eye could literally not follow," performing "piece-work," as it were, one repeated action throughout the entire day. Sinclair writes that such workers are "apt to have a family to keep alive"; moreover, on the individual level "stern and ruthless economic laws had arranged it that she could only do this by working just as she did, with all her soul upon her work, and with never an instant for a glance at the well-dressed ladies and gentlemen who came to stare at her, as at some wild beast in a menagerie."[23] These voyeuristic tourists of the plant, like the newly arrived Rudkuses, move along when they observe the "two wrinkles graven in the forehead, and the ghastly pallor of the cheeks," but Sinclair is careful to observe that the factory workers do not have the privilege to move away from the scene when it becomes tedious.[24] Sinclair dramatizes this scene, as it is the well-dressed tourists' unintended awareness of their shared humanity that makes them feel compelled to awkwardly move along. At the same time, he demonstrates that the women being observed are now somewhat comparable to the cattle seen earlier; a woman becomes a curiosity, a "beast in a menagerie," her humanity slowly being stripped from her. In another parallel between human and beast within a system that abuses both,

Sinclair implies that the woman's pallor and wrinkled face may, in fact, denote a disease obtained through the pollutants of the meatpacking plants. This job, in other words, is most likely killing her, more slowly than but just as surely as it kills the livestock for her to process. Sinclair makes this example a brutal indictment against one of the most dramatic forms of U.S. agri-expansion.

One aspect of the novel often not given as much attention as his graphic scenes of gross negligence on the kill floor is Sinclair's study of the lack of human connectivity when people are struggling to survive. In fact, he specifically discusses ethnic differences in the novel, and how the Beef Trust turns one ethnic minority group against another as a way to keep everyone's wages low. During a strike, the Beef Trust turns elsewhere for scabs—replacement workers: "these specimens of the new American hero contained an assortment of the criminals and thugs of the city, besides negroes and the lowest foreigners—Greeks, Roumanians, Sicilians, and Slovaks. They had been attracted more by the prospect of disorder than by the big wages."[25] His terms are all negative in this passage, "thugs," "criminals," and "lowest." As is seen with the Rudkuses through the targeted and misleading advertisements sent to them in Lithuania, as well as to other nations, the Beef Trust also specifically targeted rural African Americans in the South to compete against them. In other words, the Beef Trust sought to manipulate groups of people who could not know the conditions of the Union stockyards firsthand. Sinclair shows how one group of oppressed people can be quick to harm another. His narrator declares: "The ancestors of these black people had been savages in Africa; and since then they had been chattel slaves, or had been held down by a community ruled by the traditions of slavery. Now for the first time they were free,—free to gratify every passion, free to wreck themselves."[26] Sinclair is particularly harsh in his descriptions of African American characters here. However, he is honest in his portrayal of the manipulation acted upon them by the Beef Trust. Readers are able to bring a sympathetic reading to them in the same way that we are encouraged to with the Rudkuses. None of the workers are the villain in this tale; all are literally and figuratively struggling to survive. It is the Beef Trust that is a danger to all workers, as Sinclair repeatedly observes.

To some extent this is shown in the slaughterhouse tour scene. At this early stage of the story, African American and Euro-American workers are seen as relatively equal within the world of the stockyards. As one worker could be studied as if in a menagerie, in another scene, men of various nationalities and ethnicities are observed working together: "Upon both sides of this wheel there was a narrow space, into which came the hogs at the end of their journey; in the midst of them stood a great burly negro, bare-armed and bare-chested. He was resting for the moment, for the wheel had stopped while men were cleaning up."[27] When there is no stress on the system, different ethnic groups manage to exist together in relative peace. One man rests while others do their part on the kill floor. It is when a strike places stress on an already precarious system that the strain becomes too much for anyone to thrive within. Then, too many people are seeking too few jobs, and prejudices grow more pronounced. None of the characters—with the brief exception of Jurgis—benefits from these prejudicial practices, though. Only the Beef Trust thrives at this time of hardship for all the workers. In addition to drawing attention to the various ethnicities and their place in the Union stockyards, Sinclair also notes the range of cities that are comparable to Chicago through the cattle industry. He shows that these packinghouses and connected Beef Trusts exist throughout the booming cities of the Midwest—and even to the coasts. What occurs in Chicago, he maintains, is not limited to Chicago. When characters prepare to strike, they send word to "all the big packing centers,—to St. Paul, South Omaha, Sioux

City, St. Joseph, Kansas City, East St. Louis, and New York."[28] These cities will be mentioned again in Eaton's *Cattle*, discussed later in this chapter.

Sinclair's famous statement about hearts and stomachs bears other relevance in *The Jungle*, according to J. Michael Duvall. He notes that Sinclair "was following a venerable recipe for foment-ing moral judgment: begin with your basic jeremiad, ladle in lib-eral amounts of the filthy and the revolting, and stir."[29] Eventually, Sinclair imagines a socialist ending to industrial agriculture as the best solution to improve the lives of the largely immigrant populations discussed in his story. He moves, interestingly, far-ther away from the small-scale agrarianism lauded by the nation's founding fathers and the Coopers; instead, he turns toward the ideal of even further intensified agriculture as his ideal solu-tion to the Beef Trust. A socialist leader speaks to the crowd: "I was brought up on a farm, and I know the awful deadliness of farm-work; and I like to picture it all as it will be after the revolution. . . . apples and oranges picked by machinery, cows milked by electricity—things which are already done, as you may know. . . . And to contrast all this with our present agonizing sys-tem of independent small farming."[30] Ironically, Sinclair hopes to help the situation for the Rudkuses and other immigrants within the cruelty of the system of domestic animal production in the United States through socialism—and an almost science-fiction style of mechanized farming, more industrial yet than the one he critiques in *The Jungle*. Indeed, Sinclair frankly insults those who work on small-scale farms. He describes a farmer as "scratching the soil with its primitive tools, and shut out from all knowledge and hope, from all the benefits of science and invention, and all the joys of the spirit—held to a bare existence by competition in labor, and boasting of his freedom because he is too blind to see his chains!"[31] While Sinclair has spent almost the entirety of his novel lamenting the circumstances of large-scale food production in the United States, it is clear that his solution is not a return to small-scale farms in the agrarian model. He argues that such agricultural laborers are no better off than the Rudkuses and the other characters in his novel. In this part of a longer speech on the benefits of socialism, Sinclair shows a localization of agriculture at the level of U.S. meat consumption. Appreciably, Eaton makes the opposite argument in her Canadian-based novel *Cattle*, where large-scale ranchers dominate and the nobler small-scale farmers are perpetually threatened but ultimately victorious.

Global Cattle

Although a Canadian of Chinese and British descent and a younger sister to Edith Maude Eaton, who is known for her writings on the Chinese American experience in the United States under the pseudonym Sui Sin Far, Winnifred Eaton Babcock Reeve made a name for herself—Onoto Watanna—writing Japanese novels for a mostly U.S. readership.[32] While recent critics have tended to focus on her racial "tricksterism" and the issue of her "passing" in her writing, I am interested in examining Eaton's use of the materiality of food, specifically beef and beef cattle, as an opportunity to discuss identity—in terms of race and class and their connection to the natural environment. Examining a slice of Eaton's writing—focusing in this case on the phase of her professional life in which she wrote of her cattle-ranching experience in Canada—the theme of food develops greater significance as she and the nation's attitudes toward the Chinese and the Japanese change. Eaton never became an activist like her sister, but in *Cattle* she does undergo a shift in tone (see fig. 18). After writing nine successful Japanese novels, most often highlighting romances between U.S. American men and Eurasian women, she returns to her Canadian roots in *Cattle*.[33]

By setting the novel in Alberta, the Great Plains region of Canada, Eaton returns to the region discussed in the first half of the

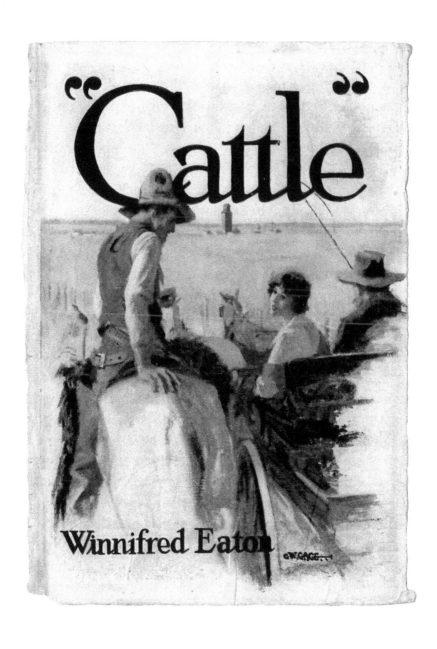

Fig. 18. George Gage's illustration for the cover of Winnifred Eaton's
Cattle, 1924. Library of Congress, Prints & Photographs Division.

nineteenth century by Washington Irving and James Fenimore Cooper, though across the national border. *Cattle* involves a small cast of farm families drawn together by their work and violence. The heroine, Nettie Day, is a humble farmer's daughter. The villain, Bill "Bull" Langdon, is the richest rancher in the area. He manipulates events to force Nettie to work for him (technically, to work for his saintly and ailing wife). While she lives with him, he rapes her and she becomes pregnant. Feeling unworthy of her intended, the neighboring farmer Cyril Stanley, she seeks to harm herself, even contemplating suicide. She is eventually healed with the help of Angella Loring, the former cook at the Langdon ranch and current hermit, and Dr. Angus McDermott, a kindly local doctor. Eventually, the villainous character is ruined—during a natural disaster—and the small-scale farmers survive and begin to rebuild. The dramatic action of the novel occurs entirely around the cattle ranches and farms of Alberta, and the natural environment is often as much of a character as the human. Eaton, through the character Dr. McDermott, theorizes, "To farm is to gamble on the largest scale possible. . . . But, like all gamblers, we are reaching out for a prize that enthralls and lures us, and that 'pot of gold at the end of our rainbow' is the harvest—the wonderful, glorious, golden harvest of Alberta."[34] In this instance, of course, McDermott means the metaphor of wealth from the harvest—economic wealth and the immediate bounty of foodstuffs—and the golden color of the wheat and corn crops, grown to feed humans and grazing animals alike. While cattle are not specifically mentioned in this passage, they are implied in the discussion of livestock feed as "gold." Cattle are a preeminent presence throughout the novel, from Langdon's prize bulls to the dairy cows on Loring's small homestead.

Bull Langdon is clearly and unsubtly the novel's villain. Eaton does not concern herself with much deep character development in the novel. It is less clear, though, if she means for his villainy to be connected to his greatness as the largest and wealthiest cattle rancher in the region—in other words, does she consider his wealth and villainy as equated somehow? She does establish that he originally earned his money immorally, suggesting that his villainy is an essential condition in either case: "For many years the Bar Q cattle had had the right of way over the Indian lands, the agents who came and went having found it more profitable to work in the interests of the cowman than in those of mere Indians."[35] Eaton returns to the arguments made by Sarah Winnemucca about abuses made against indigenous communities by immoral agents and cattle ranchers in the West. Langdon is one of these "rustlers," stealing cattle from the Indian lands themselves.[36] Ironically, this theft actually leads to his sense of pride. He considers himself the success story of the region, while the other people living there tolerate and fear him but do not like him. Langdon prides himself with his idea of success, though: "Somewhere, somehow, the Bull had come upon a phrase of the early days that appealed vastly to his greedy and vain imagination. 'The cattle on a thousand hills are mine!'"[37] Once again in *Cattle Country*, this biblical reference appears in a work about cattle. In this case, Langdon applies it to himself, rather than having someone notice a natural formation or arrangement. Eaton takes Langdon's misanthropy one step further: "To him cattle and men were much alike. Most men, he asserted, were 'scrub' stock, and would come tamely and submissively before the branding iron. Very few were spirited and thoroughbred. . . . If the Bull looked upon men in the same way as on cattle, he had still less respect for the female of the human species."[38] This passage contains a good deal of foreshadowing. First, readers get a glimpse of what awaits Nettie at his hands. In addition, we see an equation of humans with nonhuman animals. Here, Bull, ironically nick-named, is the one who believes that humans and animals are to be equated, and in this way he dishonors both. By the end of the novel, Eaton will bring animal and human together in a literal and metaphoric way. Throughout, Langdon shows a disrespect of the land, the First Nations and Euro-Canadian people who live on that land, and animals that causes Eaton's readers to wish him to be corrected. We long for justice and want to see him punished, not only for his treatment of the First Nations people off the page, or Nettie brutally within the novel itself, but also for his treatment of the land and its animals in his exploitative form of agriculture. In this way, he is part of the exploitative agri-expansion exemplified by the Beef Trust in Sinclair, also villainous in their treatment of animals, humans, and the land.

As a way to prove the value of his stock, Langdon brings Cyril Stanley with him on a tour of the major U.S. stockyard cities. Eaton writes: "Want chu to git them bulls in shape for the circuit. Goin' to exhibit in St. Louis, Kansas City, Chicago, San Francisco, and other cities in the States."[39] Eaton herself makes the connection to the real-world matter of the Chicago stockyards (and their equivalents in other cities primarily in the U.S. Midwest). Langdon is "flattered" to see "his name stamped upon the beef that topped the market, not merely in the east but in the west, even into the Chicago stockyards—there to be exhibited."[40] In the cattle-related plot of the novel, Eaton makes a pointed comment about the industrialization of agriculture in the United States, witnessed from a Canadian's point of view, a more global perspective. Once again, readers are not sure what Eaton means for us to believe with this comment coming from Langdon's perspective. When Eaton describes the environment of agricultural Alberta for her readers, she uses McDermott as her voice. Langdon is often the character she uses to show the worst form of agri-expansion in Canada in the early twentieth century. If, then, he is flattered that he bears a connection to U.S. agricultural practices, it is most likely not Eaton's own belief being described here, even as a cattle rancher herself.

The details of Eaton's writings, specifically her descriptions of Canadian cattle production and its connection to the global cattle market dominated by the United States, are useful sites in which to examine questions about culture and identity—specifically regarding the history of the Chinese in the United States during her time living and working there. In the United States, virulent anti-Chinese sentiment prevailed into the twentieth century. This xenophobia, however, does not seem to have extended to Chinese cuisine. Indeed, Chinese food is unusual in its sustained popularity throughout its history in North America. It became one of the most popular ethnic food choices in the nation as early as the mid-nineteenth century, and its popularity has never faded. In the middle of the nineteenth century, tens of thousands of Chinese—almost entirely men—emigrated from four counties of the Guangdong province of China to California and neighboring regions of North America, including Canada.[41] Living mostly apart from the Euro-American and Canadian population centers, in areas that were called Chinatowns, and performing the work that would have been considered "women's work" on the mostly male frontier, like cooking and cleaning, Chinese immigrants found ways to make do.[42]

An example of this type of Chinese cook character appears in *Cattle* in the figure of Chum Lee. Eaton names him but does not necessarily present him sympathetically in the novel. He is almost a foil for Angella Loring's character. Lee now runs the "cook car," previously run by Loring, feeding the workers on the Bar Q ranch: "Now a thin and musty smelling Chinaman dominated the car, a shrinking, silent figure, who banged down the chow before the men, and paid no heed to protest or squabble, save when the 'boss' came in, when Chum Lee became frenziedly busy."[43] While this is a decidedly negative portrait, Lee is described more sympathetically later, through his nostalgia for home. Furthermore, he suffers during a massive plague that passes through the region, killing almost half of the secondary characters of the novel and affecting everyone. Lee witnesses the death of a bunkmate, a secondary character named Batt Leeson, and becomes terrified about what will happen to him. He knows he has no one to turn to for sympathy in Bar Q or in the rest of Alberta, seemingly. He "sent up frantic appeals to the gods of his ancestors" to help him from "the dreaded curse which had befallen the land in which he had sojourned too long."[44] In this attitude, he decides to perform one last act of kindness: he sets Bull Langdon's starving Hereford bulls free.

In this way, Eaton allows her Chinese character, Lee, to perform an act that influences the plot of the rest of the novel, in a way reminiscent of María Amparo Ruiz de Burton's character Chapo. Moreover, he performs an act of kindness almost as great as that of anyone else described, including Loring and McDermott, one that ultimately harms Langdon and helps the victims of his cruelty. Interestingly, though, his act is seen as ambiguous in the novel at the time he performs it. He is acting out of kindness, but also because he wants to appease his gods. Moreover, he has given the bulls a chance to survive in the unfenced wildness, but it is unclear how good their chances of survival are. Fenced in with only dying hands to feed them, the bulls are most likely going to starve to death; however, Lee releases them into the Alberta winter to forage for themselves. Eaton writes: "Pampered and petted, used to being fed almost by hand, and knowing no range save the sweet home pastures, how were they likely to fare in the wilderness? Now the merciless cold of the implacable winter smote them to the bone, and the unbroken expanse of frozen snow rose four feet deep in mounds and hillocks on all sides of them."[45] At least, though, there is a chance that these bulls may live following Lee's act; overall, therefore, Lee acts well. Moreover, Prince Perfection, Langdon's prizewinner, can survive to perform one remarkable deed by the end of the novel (see fig. 19). Eaton gives her Chinese character an important secondary role, and in so doing she demonstrates a significant fact about cattle ranchers of the United States and Canada during this time. Namely, they were not all of European ancestry; rather, they were of a variety of ethnicities. In the nineteenth century, U.S. businesses actively recruited Asian immigrant labor for railroads, agriculture, and other economies.[46] The agricultural economy included, obviously, cattle ranches and grain farms throughout the Great Plains, in the United States and Canada. As Sinclair shows the various ethnic groups laboring in the stockyards of Chicago at the turn of the century, so Eaton shows some of the ethnic diversity on the farms and ranches themselves. Jill Lepore observes that by 1865 the railroad had already made large-scale beef cattle a booming industry. The cowboys working on these ranges were from a range of races, nationalities, and ethnicities, and the Chinese were working alongside all of them.[47]

In a kind of posthuman cowboy justice, Bull Langdon is finally stopped by being gored to death by Prince Perfection after the ranch had been neglected due to Langdon's single-minded quest to reclaim Nettie after she escapes from him. "Again and again the savage bull gored and tossed him until he was rent into pieces," Eaton writes. "A master vengeance was in that act of justice, though no torture of Bull Langdon's body could atone for the torture he had inflicted upon Nettie Day's soul."[48] Eaton enjoys the somewhat heavy-handed irony of the bull destroying Bull, the novel's nearly eponymous villain. When Prince Perfection is set free by Lee, he does not accidentally gore Nettie or Cyril, for example, but rather the one human deserving of death in the novel. In this way, Eaton adds a moral element to her story, as well as an ecocritical one: nature defeats human hubris in the end. Cattle are shown to be one of the most important symbols throughout the novel. Indeed, in the final scene, cattle work figures prominently, as Nettie is shown to be working with her dairy cattle

Fig. 19. William H. Howe's painting *Monarch of the Farm* (1891) gives a sense of the grandeur of a prize bull like Prince Perfection. Smithsonian American Art Museum, gift of Mrs. William Henry Howe.

when Cyril comes upon her. She meets him with "a pail of milk in either hand," as they have their culminating romantic moment together.[49] The pairings of Nettie and Cyril and of Angella Loring and Angus McDermott present them as the sentimental victorious survivors, living on their relatively small farms when compared to Langdon's, demonstrating a more appropriate respect for the natural world—a dramatically different conclusion than Sinclair's proposed industrialized agriculture through socialism described at the end of *The Jungle*. Eaton's ending, though, brings North American agri-expansion narratives nearly full circle, to the small-scale agrarianism extolled by earlier writers.

Rice Is the Answer

What, then, is the appropriate form of agriculture, according to Eaton or Sinclair? What is good to eat? Following a shift in public opinion regarding Japan, Eaton began to alter her subject matter to some extent in the 1910s. Her writing went through a transformation when she collaborated with another of her sisters, Sara Eaton Bosse, in their *Chinese-Japanese Cook Book* (1914). Consumers were starting to engage in a touristic form of cultural consumption through travel narratives at this time, and in many cases this literary tourism was followed by an interest in the cuisines of those same nations and regions. There was a growing national interest following the turn of the century, therefore, in ethnic cuisines, and Eaton recognized a market for that kind of cultural integration, broadening U.S. palates and culture to include Chinese as well as Japanese cuisines. Laura Shapiro observes that by 1913, less than a decade after anti-Chinese immigration legislation had been extended indefinitely, "livelier tastes and traditions that still flourished in immigrant families and communities were supplying the nation with more adventurous perspectives on food."[50] This was an exciting period in the United States in terms of ethnic and regional cuisines, Chinese food being one of the most striking examples. In "Recipes for Reading," her classic study of the *Joy of Cooking*, Susan Leonardi argues that the narrative element of a cookbook is a critical aspect of its success. The stylistic details matter in "an embedded discourse."[51] In the *Chinese-Japanese Cook Book* the placement of "Chinese" before "Japanese" in the title and within the cookbook itself places Chinese cuisine literally and culturally before Japanese cuisine. While the nation enacted repeated severe Chinese exclusion acts, it was developing a passion for Chinese dishes, including the early fusion dish of chop suey. Chinese cookbooks—of which the *Chinese-Japanese Cook Book* is one of the earliest examples—became a potential counter against xenophobia during this period.[52]

The traditional Chinese and Japanese staple, rice, is privileged throughout the cookbook. Instructions regarding its proper preparation are given in both the preface and the "Chinese Recipes" section under the heading "To Boil Rice." First, the sisters declare, "Rice is indispensable."[53] The emphasis in the cookbook's preface shows the great significance of rice: "The secret of the solid, flaky, almost dry, yet thoroughly cooked rice lies in the fact that it is never boiled more than thirty minutes, is covered twenty minutes, never stirred nor disturbed, and set to dry on back of range when cooked, covered with a cloth. Mushy, wet, slimy, overcooked rice is unknown to the Chinese and Japanese."[54] Bosse and Eaton, as the cookbook's primary cook and writer, respectively, attempt to explain the difference between properly and improperly cooked rice. Rice should be "flaky," "almost dry," and of course "thoroughly cooked." Furthermore, they compare rice to potatoes and bread as starchy complements to the protein, usually beef or another form of meat, of a U.S. meal. When Bosse and Eaton instruct their readers on the importance of avoiding the "mushy, wet, slimy" rice, they are defending Chinese and Japanese cuisines against detractors who may have suffered a negative experience through poor preparation. Bosse and Eaton are therefore attempting to provide a cultural studies course in miniature through this early passage of their collaboration by teaching U.S. cooks how to properly prepare rice. Indeed, to some extent, Eaton seems to return to one of the themes of *Tama*— before Tojin arrived with his excessive meat requirements—in this section promoting the healthfulness of rice consumption.

Sinclair also extols the benefits of rice in his dietary writings. His unpublished essay "My Life in Diet" shares much with his work published in *The Fasting Cure* (1911) and *The Autobiography of Upton Sinclair* (1962). His wife, Craig, had an enlarged heart and sought a heart-healthy diet. After visiting several sanitariums, including the famous "San" run by John Harvey Kellogg in Battle Creek, Michigan, they began a "rice diet" and discovered it to be the most successful. Building upon Craig's health concerns, Sinclair is able to recommend a rice-focused diet from personal experience:

> So there I was, doctor's apprentice, nurse, cook, and bottle-washer, for several months. She was living on rice, fruit, and fruit juices, and vitamins—and doing so well that it was natural for me, life-long health crank, to announce: "I'm going to cook an extra pot of rice for myself and see how it works." Craig was afraid, but I went ahead, and my fate was decided in a few days; call me a Hindoo, call me a Jap, call me a Chinaman, call me anything you please—rice is the food for me. Just think of it: I, who for at least forty years had been accustomed to saying that I was never more than twenty-four hours ahead of a headache, have not had a headache for six and one-half years; and I am quietly confident that I shall never have another as long as I live.[55]

Interestingly, Sinclair writes that his consumption of rice has made him something like "a Hindoo," or even "a Chinaman," terms that would have been in use during his time. In this way, he practices some of the same essentialization that Eaton does in *Tama*. Moreover, he is as determined in the truth of his health through the "rice diet" as he was about the injustices of the Beef Trust in *The Jungle*. Rice works for Craig and himself, and he wants to encourage others—largely mainstream Euro-American consumers—to try this diet as well for their own health. A severe diet, it does not allow for dining in restaurants much. However, Sinclair approves of rice as "the food for me" because it gives his family back their health.

Almost the diametrically opposite form of agriculture from cattle husbandry, rice came to North America—where both Eaton and Sinclair would be able to enjoy it—from the African trade, European colonization, and Asian immigration. Francesca Bray writes that, to the high-end "Spice Road" trade goods like "spices, rare woods, and costly dyestuffs, exquisite printed cottons, lustrous silks, and porcelain," were added "traffic in mundane necessities" like rice.[56] From those beginnings, rice came to the land that would become the United States, developing part of the story of this settler colony. While John Winthrop's seventeenth-century Puritans were trading cattle to the Caribbean, the colonists of the Carolinas were using the African slave system to help with growing rice for the Caribbean trade, as well. During this early period, then, rice was, like cattle, already embroiled in global economies. In the nineteenth century, the century of expansion and cattle, rice production also developed throughout Louisiana and California, largely connected to Chinese immigrants and their influence. Rice is not used much as feed for cattle, ironically, compared to other major cereal crops like corn and wheat. Thus, as Eaton and Sinclair discuss rice as an alternative to beef, they are going from one culinary extreme to another. Furthermore, the Asian influence of rice is emphasized in both writings, although rice had also been domesticated in Africa and the Middle East—as well as Europe—for hundreds of years. Nevertheless, both authors essentialize the foodstuff with one particular geographic region: East Asia. It is interesting, then, that they argue for foods from Asia becoming the secret to delicious cuisine and good health. Perhaps more work could be done on the influence of rice in the literature and culture of the United States. Regardless of the advice of Sinclair and Eaton, however, beef remained the choice diet of the United States. Indeed, over the twentieth and twenty-first centuries, consumers continued to eat ever greater amounts of beef at the national and, increasingly, global levels.

Conclusion

Meat Is the Message

Pork is possible, but beef is best.

—RUTH OZEKI, *My Year of Meats*

At the end of the twentieth century, Ruth L. Ozeki described a globalizing cattle industry in *My Year of Meats* (1998). Written almost a hundred years after Upton Sinclair's *The Jungle*, it shares with its predecessor a sense of muckraking intensity about some of the dangers of the intensive meat production industry in the United States. Although Sinclair's book helped bring about needed regulations, much about the industry has remained the same. Ozeki's novel predates related nonfiction works that engage with the continued issue of intensive meat production and the beef industry, such as Eric Schlosser's *Fast Food Nation* (2001) and Michael Pollan's *Omnivore's Dilemma* (2006). In addition, Ozeki continues some of the cosmopolitanism and questioning of essentialization based on food identities seen in Winnifred Eaton's novels. Thus, she makes a timely modern synthesis of their work. Ozeki's book follows two protagonists: Jane Tagaki-Little is a biracial U.S. filmmaker who gets a job working for a Japan-based television show called *My American Wife!*; meanwhile, Akiko Ueno is her Japanese counterpart, an abused wife married to one of the show's producers. The intercalated point-of-view chapters deal with their growing concerns regarding the show, ostensibly designed to sell more beef to a Japanese market. Over the course of the novel, Tagaki-Little begins to surreptitiously subvert that design by showing some of the hazards of the U.S. meat production industry. Eventually, Ueno finds her own independence through her relationship to the show and Tagaki-Little. Throughout the novel, Ozeki challenges the nation's—and to some extent, the world's—ongoing drive to make meat cheaper and more plentiful to an increasing global population, seemingly regardless of the dangers to the human and nonhuman world that entails. She observes, somewhat tongue in cheek, that "cheap meat is an inalienable right in the U.S.A., an integral component of the American dream."[1] This right is one that large corporations want to expand to new nations, specifically Japan and more generally the Asian continent. The novel's climax occurs at a Colorado feedlot, in the heart of the U.S. West, in a series of events that involves a discontinued hormone called DES, issues of fertility and cancer in animals and humans, and the injury and hospitalization of both women in their respective regions. In her seriocomic muckraking novel, Ozeki follows the methodology of her nineteenth-century and early twentieth-century literary counterparts, using the symbol of cattle as representative of the United States (see fig. 20).

The agri-expansion discussed in *Cattle Country* changed over the twentieth century and the closing of the physical U.S. frontier, but it did not entirely go away. Indeed, it intensified, transforming into what we now call globalization during the twentieth century, the "American Century," and it continues unabated into the twenty-first century. This transnational move has been felt perhaps in no industry more profoundly than that of industrial meat production. In the twentieth century, hormones, antibiotics, and food additives helped make the production of beef cattle even more efficient than nineteenth-century advances had. Gone were

Fig. 20. Photo from a 2019 visit to the experimental farm at the University of California, Davis. Author's collection.

the open ranges, replaced by feedlot cattle. Also gone were the Great Plains as the dominant region for beef production, replaced by an ever-expanding series of nations, including prominently Brazil and Australia. The United States is not, in fact, the only or even the greatest producer of beef or cattle in the twenty-first century. As of 2019 it was the fourth-largest beef-exporting nation in metric tons, after Brazil, India, and Australia, with China making significant gains as well.[2] In addition, the United States is not the leading consumer of beef, either, as China now eats more beef, with India, Brazil, and Mexico rounding out the leading importers.[3] The Argentinian gaucho/a and Australian jackaroo/jillaroo figures are now growing nearly as iconic as the historic U.S. cowboy/girl, and the "cowboy kingdom" that developed in the 1880s in the United States is quickly spreading across the globe. Simultaneously, the environment is being affected more dramatically than ever in the twenty-first century, thanks in large part to global meat production practices. In 2019–20, in fact, the Earth was literally on fire in regions known for their relation to the cattle industry, such as Brazil, Australia, and the western United States.[4] In *Cattle Country*, I have focused on nineteenth-century narratives that dealt with cattle specifically. Interestingly, the works discussed in this book predate the "green revolution" following World War II and modern globalization—what I have been calling "globalization" in the nineteenth century is truly "proto," leading as it did to the hyper-industrialization and multinational corporations of the late twentieth and early twenty-first centuries as described by a new generation of authors like Ozeki.

I have argued that although cattle have been an integral part of U.S. Manifest Destiny and a harmful agri-expansionism, they also can be used to show a transregional and posthuman awareness, as seen in the works discussed throughout *Cattle Country*. Specifically, I have shown that the authors of the long nineteenth century repeatedly chose cattle as the animal that most represented the United States during agri-expansion. Cattle are uniquely suited to this role. They are paradoxical animals: although they must have a particular living environment, their human companions have rearranged lands and peoples in order to provide them ample food. Cows are actors as well as subjects being acted upon—successful beyond nearly any other animal species, and in that way directly contrasted to the bison—through their connection to the humans who raise them for their dairy, meat, and other products of their bodies. Regional and national borders across the North American continent, in fact, overlapped and intertwined throughout the nineteenth century because of this cow-based paradox. Moreover, cattle's own mobility was first symbolic of the frontier and U.S. settlers, then more tragically of the forced removals of peoples to make room for cattle ranches and monoculture agriculture. While animals might have been anthropomorphized in literary representations, I argue, people have also been dehumanized. In addition, the contrast between cattle and bison continues in the Great Plains region of the United States—the conservation of bison in national parks like Yellowstone remains a controversial issue. The contrasting of cattle and bison continues to show the importance of reimagining a dominant narrative, as with cattle and agri-expansion. In some cases, writers show positive examples of cattle as symbols of redeveloped, less-intensive agricultural systems: they are shown coexisting peaceably with small-scale farmers, representing the iconic U.S. West and its association with open spaces, and even punishing villain characters. All of the authors described in *Cattle Country*—Washington Irving, James and Susan Fenimore Cooper, Henry David Thoreau, Sarah Winnemucca Hopkins, María Amparo Ruiz de Burton, Charles Chesnutt, Upton Sinclair, and Winnifred Eaton—interrogate the use of cattle, as animal and symbol, for and against the cause of agri-expansion.

Meanwhile, Ozeki continues the critical work begun by Sinclair and Eaton in *My Year of Meats*. She observes the expansion of the beef industry across the globe, as opposed to merely across the nation. At the same time, she questions the appropriateness of that expansion. Where Sinclair names the Beef Trust of Chicago his villain, Ozeki describes a syndicate named BEEF-EX: "BEEF-EX [Beef Export and Trade Syndicate] was a national lobby organization that represented American meats of all kinds— beef, pork, lamb, goat, horse—as well as livestock producers, packers, purveyors, exporters, grain promoters, pharmaceutical companies, and agribusiness groups. They had their collective eye firmly fixed on Asia."[5] In this way, she synthesizes and modernizes the issues raised by Sinclair and Eaton, respectively. A transnational, multifaceted conglomeration seeks to expand its market through dubious means. At the same time, throughout her work, Ozeki shows red meat to have an essentializing, U.S.

nationalistic quality that can be promoted to the rest of the world. In this novelized example, beef is sold to Japanese consumers as the particularized way in which to literally incorporate the culture of the United States. To eat beef is, to some extent, to identify with the United States, even to the Japanese. Ozeki makes this suggestion more concrete through her biracial main character, Tagaki-Little, who is a tall Asian American woman, at almost six feet, her height being made much of when she spends time in Japan. She is, it is implied, what is possible if Japanese people consume more U.S. beef.

Of course, Ozeki next challenges this implied belief, showing the lie underneath the marketing. Tagaki-Little is sterile, a DES daughter; she had cancer in her early twenties, as well as reproductive complications: "DES, or diethylstilbestrol, is a man-made estrogen that was first synthesized in 1938," Ozeki writes, going on to give a brief history of its development in the cattle industry through the second half of the twentieth century, as it was introduced as an additive to cattle feed in 1954 before its carcinogenic nature was discovered. "Finally, in 1979, the government banned DES for use in livestock production. . . . Today [mid-1990s], although DES is illegal, 95 percent of feedlot cattle in the U.S. still receive some form of growth-promoting hormone or pharmaceutical in feed supplements."[6] While she assumes in the novel that her mother has been given DES during her pregnancy, in a tragic allusion to the drug Thalidomide, Tagaki-Little comes to learn that it is also in the beef she has been eating as a representative U.S. midwesterner. She could plausibly have been harmed through her iconic U.S. beef-based diet.

In the novel's climax, Ozeki describes a Colorado feedlot in a scene that could have come straight out of either Sinclair's *The Jungle* or, to some extent, Eaton's *Cattle*: "From a distance, the feedlot itself looked like an island, an enormous patchwork comprising neatly squared and concentrated beef-to-be. Angus,

Brangus, Hereford, Charolais, Limousin, and Simmental, these were breeds, not animals, penned with precision and an eye to slaughter that was antithetical to the randomness of living things."[7] The feedlots are no longer in Chicago and other large midwestern cities. Now they are located farther from urban centers, like a rural area in eastern Colorado. Importantly for Ozeki, the sameness witnessed at the Colorado feedlot is itself a lie. The "American" life being promoted is in fact not a monoculture, and has never been, regardless of the drive of its people to consume cheaper, more plentiful meat. The families within the BEEF-EX promotional videos are diverse and regional: they live in the South and the West, they are largely poor, they are also not all Euro-Americans, heteronormative, or even meat consumers. They are in their own lives contradicting and challenging the representation of the United States as one thing only, as only the land of beef. In this way, Ozeki continues the themes discussed throughout *Cattle Country*. She and her characters are being manipulated by, colluding with, and trying to resist BEEF-EX; furthermore, they are deeply concerned about the human and environmental implications of intensifying agriculture. They resonate with Ozeki's readers, as well as allow her readers to reimagine the authors discussed in *Cattle Country* as being similarly concerned with the rapid changes occurring in their own time.

Therefore, while it is challenging to apply too many theoretical approaches to a literary reading, the works analyzed in *Cattle Country* have necessitated a particularly capacious interdisciplinary approach. It turns out that the symbol of the cow, like Walt Whitman's narrator, is large and contains multitudes. Therefore, I combine food studies and agricultural studies as they describe the long overdue account of U.S. settler colonialism. As the site where culture and nature come together, food has a materiality beyond nearly all other cultural markers.[8] Donna Gabaccia argues that as food can be seen as one of the "concrete symbols" of society, "eating habits both symbolize and mark the boundaries of cultures."[9] Analyzing the changing U.S. culture during the nineteenth century has been one of the primary concerns within *Cattle Country*. I have been influenced by animal studies and ecocriticism—the work of Lawrence Buell, Donna Haraway, and Cary Wolfe, among others—in the way these lenses help to engage with nonhumans and the land, what is now called the environment. I also employ a historical approach to the long nineteenth century in the form of its study of grass cultivation in the United States, in the guise of cattle production, how beef gained its centrality in the U.S. diet over that time.[10] Moreover, I engage with critical race studies of food, focusing on the work of Catherine Keyser and Kyla Wazana Tompkins as such work engages specifically with literature and nature, food, and issues of sovereignty, especially when contrasting the figure of the cow moving ever westward across the continent with the people who are dispossessed by that movement.

Finally, it seems as if I cannot complete the editing of *Cattle Country* in 2020 without commenting on the pandemic that was occurring during this time, COVID-19. The worst pandemic in a hundred years, it was being compared to the Spanish influenza outbreak of 1918 and 1919. I asked myself in the spring of 2020, how does this pandemic connect to a narrative history of cattle in U.S. and global culture? A brief search of online news venues showed numerous headlines about meat and COVID-19. For example, I saw in *The Guardian* that Paul McCartney was calling for the Chinese wet markets to be banned, based on unconfirmed reports that these markets were the epicenter of the virus. The fear surrounding the virus emphasized ongoing issues including food security, the danger of xenophobia, and globalization and its effects on regional communities. Returning to the concerns raised in *My Year of Meats*, a CNN article indicated that a Colorado meatpacking plant closed for a week in mid-April to test employees for the virus and to do a deep cleaning of their facility. This brief story was in response to a much larger matter. An *LA Times* article reported that hundreds of workers at Cargill meatpacking plants—including ones in Colorado—had tested positive for the virus. COVID-19 had been found in people working in other agricultural companies like JBS and Smithfield, as well, and some have even died of the virus. Issues of food security, human safety, and "the fragility of the food-supply chain" were being exposed and deeply felt during this time of fear and isolation caused by COVID-19. People went through phases of hoarding food and household items, as well as showing increased concern related to the threat to national and international meat supplies. Meat, in particular, became temporarily rare in grocery stores. Ironically, Kim Cordova, Local 7 of the United Food and Commercial Workers Union president, told reporters for the *LA Times* that "you cannot make sacrifices like this with people's lives. . . . People can live without beef."[11] This issue was ongoing as this book went to press. The subject of *Cattle Country*, cattle, was again shown to be of concern in terms of food sovereignty, monocultural agriculture, the effects of human-caused climate change, and U.S. expansionism that has turned into globalization. Indeed, it might have been perhaps even more essential in 2020.

In a way that speaks to both the cattle based agri-expansion of *Cattle Country* and the COVID-19 pandemic, then, in 2008, nutritionist Marion Nestle warned of the dangers of an unregulated global food system in her book *Pet Food Politics: The Chihuahua in the Coal Mine.* The book's focus was a pet food recall in 2007, after it was discovered that the Canadian company Menu Foods had bought melamine-tainted flour from a factory in China, resulting in the poisoning and death of hundreds of animals—mostly small dogs and cats. Nestle's larger argument, though, was focused on humans. The recall, she writes, "exposed catastrophic weaknesses in global safety systems."[12] This globalizing economy is notably under threat over ten years after Nestle's warning, which seems prescient at a time of global pandemic and the concern of a global financial crisis to follow. I argue that at a time like this, writers, artists, and cultural commentators are especially needed in order to help create new narratives and encourage original ways of engagement with the human and nonhuman world. Writers and others can frame conversations, challenge the status quo, and ask important questions in their work, such as How are current issues like global pandemics influencing cultural representations of cattle? Would the United States appear different if the nation's iconic animal was a chicken or a sheep, rather than a cow? What would happen to the nation's cowboy image if that figure was, instead, a shepherd? The 2020s are showing us that the world is dramatically changing, potentially into something wholly different. What will that new world look like?

NOTES

Introduction

1. Lewis and Clark, *Journals*, 476.
2. William Cronon notes that the domestication of draft animals like oxen "was itself one of the great chapters in the global history of technology" (*Nature's Metropolis*, 98). Furthermore, Diana L. Ahmad discusses cattle and oxen specifically in *Success Depends on the Animals*.
3. For a discussion of the concept of agri-expansion, see Dolan, *Beyond the Fruited Plain*.
4. Anderson notes, "These people and animals shaped the course of colonial history because of their interactions, not their separation from one another" (*Creatures of Empire*, 4). Describing modern agriculture in a way that applies as well to a discussion of nineteenth-century narratives, Donald Klingborg observes that husbandry "implies a moral responsibility on the part of the owner to care and provide for the animals under their care" (Introduction, 4).
5. Throughout *Cattle Country*, I describe a number of Native American tribes and nations. When possible, I use the name of the indigenous community. Otherwise, I use the term "Native American" or "indigenous communities" for more general discussion—using the term "Indian" only in quotations.
6. Lewis and Clark, *Journals*, 322–23.
7. Andrea Smalley observes what Lewis and Clark foreshadow in their journals: "Indigenous animals more often stood in the way of Anglo-American plans to master and possess the land" (*Wild by Nature*, 3).

8. Clark doesn't note this in his journal, but the death of the Arikara chief resulted in hostile relations between the United States and the Arikara from that point.

9. Lewis and Clark, *Journals*, 472–73.

10. Horsman, *Race and Manifest Destiny*, 107–8.

11. Outka, *Race and Nature*, 3.

12. See Howe, "Tribalography"; and Omi and Winant, *Racial Formation*.

13. See especially Kolodny, *The Lay of the Land*; Allen, "The Sacred Hoop"; and Buell, *The Environmental Imagination*.

14. Fernández-Armesto, *Near a Thousand Tables*, 220, xii.

15. Korsmeyer, *Making Sense of Taste*, 186–89. Foundational works in food studies include Douglas, "Deciphering a Meal"; Bourdieu, *Distinction*; and Appadurai, "How to Make a National Cuisine."

16. See Mead, "The Problem of Changing Food Habits."

17. Indeed, there is another element of the word that is implied, though I don't engage with it directly in *Cattle Country*: that of its connection to "chattel" as in "chattel slavery." Sometimes the word *cattle* was used to refer to enslaved peoples, an example being in Harriet Beecher Stowe's *Uncle Tom's Cabin* (1852).

18. Nelson, "Rustling Thoreau's Cattle," 262. Discussing the united history of cattle and humanity, Rimas and Fraser observe, "Imagine our world without cattle, and you're not imagining our world. Cattle, second only to the ingenuity of humanity itself, built this astounding complexity of fields and cities, letters and money, banks and kings" (*Beef*, 9).

19. Sumner, "Economics of Global Cattle Industries," 17.

20. "Bison," National Park Service, last modified February 21, 2020, https://www.nps.gov/yell/learn/nature/bison.htm.

21. Jefferson, *Notes on the State of Virginia*, 164–65.

22. Jefferson, *Notes on the State of Virginia*, 165.

23. George Washington—hardly the small-scale landholder idealized in the literature—also kept cows: "Washington's cattle were the Devon milk breed, and he appreciated their versatility—providing not only beef and veal but also milk and its derivative products cream, butter, and cheese. The herd, branded with 'GW' on their flanks, generally numbered about 150, and an additional 26 or so were draft oxen used to work plows and wagons and thus were not slaughtered for food" (DeWitt, *The Founding Foodies*, 77). Also, Amy J. Fitzgerald observes

that "a significant shift took place between approximately 1815 and 1880: agriculture adopted the characteristics of the manufacturing industry and became commercialized"; in other words, shortly after the Louisiana Purchase the United States entered a period where the nation moved from "subsistence livestock farming" to "intensive livestock farming." She also notes the "five main characteristics of industrial agriculture" as being automation, intensification, specialization, science and technology, and increase of scale (*Animals as Food*, 9, xiii).

24. See Saunt, *Unworthy Republic*.
25. Deloria, *Playing Indian*, 63–64; Myers, *Converging Stories*, 2; Specht, *Red Meat Republic*, 43.
26. Mann, *1491*, 25.
27. For more reading on Native American forms of agriculture—historical and current—see the National Congress of American Indians, "Agriculture," http://www.ncai.org/policy-issues/land-natural-resources/agriculture.
28. For a discussion of the "noble savage" history of the Boy Scouts of America, see Deloria, *Playing Indian*, 106–11. Angela Calcaterra works to reorient readings of this trope, focusing particularly on prairie works of Washington Irving and James Fenimore Cooper as they engage with Pawnees and Osages. She argues that "critical orientations attentive to encounter and proximity draw out the complicated scenes of literary production that undergird a wide range of literary depictions of Indians in this period and beyond" (*Literary Indians*, 144).
29. Stoll adds: "A symbiotic relationship would be the best that could be hoped for, but American farmers tended to drain the body that hosted them, and the vector always pointed to fresh blood in the West" (*Larding the Lean Earth*, 31).
30. Fiskio, "Unsettling Ecocriticism," 302.
31. Both *Cattle* and *The Jungle* demonstrate the powerlessness of women—especially poor women—at the turn of the twentieth century. Sexual harassment and rape occur in both works, the beef industrialists exerting their dominance over female subordinates.

1. Cattle and Indian Territory

1. Irving, *The Sketch Book*, 275.
2. Irving, *The Sketch Book*, 280. Even in *The Sketch Book*, Irving already imagined a "Lord knows where" future frontier.

3. "Traits of Indian Character" and "Philip of Pokanoket" were published first in *Analectic Magazine*, which Irving edited, in 1814, and were edited and reprinted in the British *The Sketch Book* (1819–20).

4. Irving, *The Sketch Book*, 240–41. Here, deer are being contrasted to cattle and conflated with the Wampanoags.

5. Rubin-Dorsky, *Adrift in the Old World*, 116.

6. Hoffman, "Irving's Use of American Folklore," 429.

7. Burns, "Ineffectual Chase," 59–60.

8. Michael Rogin observes that early colonists did so, "Overlooking, for their own political and myth-making functions, extensive Indian agriculture (which had kept the first white settlers from starving)" (*Fathers and Children*, 115). Indeed, Virginia DeJohn Anderson would name livestock "necessary, if not wholly sufficient, causes for these tragic confrontations." Sometimes the Wampanoags even tortured cattle and horses, taking "aim at the most 'English' of livestock species" (*Creatures of Empire*, 232, 236).

9. Turner, *History, Frontier, and Section*, 70, 67.

10. Irving's writings about the United States, its natural lands, those who lived there, and those who were expanding into those spaces coincide with Paul Outka's timeline in which "the wilderness is a racially and culturally particular construction with intellectual and aesthetic origins in Romantic sublimity and American transcendentalism." While Outka focuses on a later period of U.S. history, the tangle of race and nature had begun by Irving's time (*Race and Nature*, 2). Turner adds: "One element favoring the rapid expansion of the rancher's frontier is the fact that in a remote country lacking transportation facilities . . . the cattle raiser could easily drive his product to market" (*History, Frontier, and Section*, 70). For more on this period of U.S. history in the West, see Knowlton, *Cattle Kingdom*.

11. Irving, *The Sketch Book*, 225.

12. Irving, *The Sketch Book*, 225.

13. For analysis of U.S. policy along the developing frontier, especially regarding attitudes toward Native Americans, see Limerick, *The Legacy of Conquest*; Slotkin, *Regeneration through Violence*; and Vizenor, *Manifest Manners*.

14. Irving, *The Sketch Book*, 240.

15. Laura Murray argues that Native American cultures are described in Irving's work as part of this imperial project; his Indian sketches are

bookended by some of his most imperialistic sketches, "Stratford-on-Avon" and "John Bull" ("The Aesthetic of Dispossession," 215–16); Daniel Littlefield maintains: "It is clear to Irving that from the beginning American colonists failed to recognize the complexity of Indian culture" ("Washington Irving and the American Indian," 139); and Mark Niemeyer adds that narratives of Native Americans during this period often compared them to Greco-Roman figures ("From Savage to Sublime," 8–18). For a discussion of transatlantic Irving, see Hanssen, "Transnational Narrativity and Pastoralism"; also see Dippie, *The Vanishing American.*

16. Steven Petersheim considers Irving's works as they deal with this period of history, in addition to their regionality ("History and Place," 118). Michelle Sizemore adds that Irving inadvertently contributed to the "national fantasy by bringing a sense of *fait accompli* to Native Americans' extinction" ("Changing by Enchantment," 174).

17. Irving, *Crayon Miscellany*, xx.

18. Antelyes, *Tales of Adventurous Enterprise*, 96. Littlefield adds: "If Irving wanted to see the Indian in his closest approximation of the aboriginal state," he would have to travel to the western territories ("Washington Irving and the American Indian," 142).

19. For a recent analysis of the development of Indian Territory, see Saunt, *Unworthy Republic.*

20. Root and de Rochemont, *Eating in America*, 258–59.

21. Irving, *Crayon Miscellany*, xx.

22. Irving, *Crayon Miscellany*, xx. Indeed, Irving had read the three earliest Leather-stocking novels—including *The Pioneers* (1823) and *The Last of the Mohicans* (1826)—by the time of his own tour. He was also familiar with the popular poem; for a discussion of Native Americans in Bryant, see Jentz, *Seven Myths*, 86.

23. Bryant, "The Prairies," 131–32.

24. Mark K. Burns describes Irving's three reasons for traveling in the "authentic" Far West as portrayed in his text: "(1) the Native America tribes of the Great Plains, (2) prairies in their 'pristine wildness,' and (3) animals—especially buffalo—in their 'native' state" ("Ineffectual Chase," 57). In addition, see "Henry Leavitt Ellsworth," New York Public Library Digital Collections.

25. Irving, *Crayon Miscellany*, 9.

26. Irving, *Crayon Miscellany*, 9.

27. Irving, *Crayon Miscellany*, 10.

28. Irving, *Crayon Miscellany*, 19.

29. Irving, *Crayon Miscellany*, 19.

30. Irving, *Crayon Miscellany*, 19.

31. For more on this, see Cella, "Disturbing Hunting Grounds," 72, 65.

32. Irving, *Crayon Miscellany*, 82–83.

33. Irving, *Crayon Miscellany*, 83.

34. To be sure, this is a lot of meat to abandon; Root and de Rochemont comment that the male bison weighed around 2,500 pounds, with 1,250 pounds of usable meat (*Eating in America*, 202).

35. Irving, *Crayon Miscellany*, 102.

36. Irving, *Journals*, 119.

37. Irving, *Journals*, 121.

38. Root and de Rochemont, *Eating in America*, 66.

39. H. N. Smith, *Virgin Land*, 100.

40. Quoted in H. N. Smith, *Virgin Land*, 72–73.

41. Root and de Rochemont, *Eating in America*, 201–2. Also see Mann, *1491* and Isenberg, *The Destruction of the Bison*.

42. Irving, *Crayon Miscellany*, 120.

43. Irving, *Crayon Miscellany*, 121.

44. Irving, *Crayon Miscellany*, 121.

45. Irving, *Crayon Miscellany*, 121.

46. Irving, *Crayon Miscellany*, 100–101.

47. Schlueter, "Private Practices," 288.

48. Burns, "Ineffectual Chase," 70.

49. Peter Antelyes argues that Irving's style would influence and define "much of the literature of Western expansion throughout the nineteenth century" (*Tales of Adventurous Enterprise*, 46). Richard Slotkin, referencing Henry Nash Smith, adds that "the image of the wilderness east of the Mississippi changes from 'desert' to 'Garden' in a century and a half, while that of the Great Plains exhibits a similar change in less than half that time" (*Regeneration through Violence*, 9). For more on Irving in terms of a hemispheric reading, see Wingate, "Irving's Columbus."

50. Hardt and Negri, *Empire*, xii.

51. Irving, *Astoria*, 152.

52. This reading, though, "ignores the complex version of nationalism produced by the experience of commercial traffic among rival nations,"

Stephanie LeMenager argues, that was occurring in the nineteenth century along the Pacific ("Trading Stories," 684).

53. Antelyes, *Tales of Adventurous Enterprise*, 100. In addition, Slotkin notes that Irving's history "prophesied that those who settle in the Great Plains will become a 'mongrel race' of brigands" (*Regeneration through Violence*, 437).

54. Jaros, "Irving's *Astoria*," 2.

55. Jaros, "Irving's *Astoria*," 1.

56. In addition, according to historian Peter Stark, "Near the Pacific Coast lay a virtual Eden—a great valley of rich, moist, impossibly green land, perfect for farming. This was the Willamette" (*Astoria* 299).

57. Irving, *Astoria*, 136.

58. "Account of the Buffalo," 1. The article describes the animal by comparing it to cattle. It "is larger than an ox, has short black horns, with a large beard under his chin, and his head is so full of hair, that it falls over his eyes and gives him a frightful look. . . . Its head is larger than a bull's, with a very short neck; the breast is broad, and the body decreases towards the buttocks. . . . [I]t is the only species of wild cow know in North America, there being none like the European, but what were first carried over" ("Account of the Buffalo," 1).

59. Watson, "Lawless Intervals," 6.

60. Wayne Kime compares Irving's depiction of Bonneville the character as an idealized version of Irving himself as traveler: "The portrayed Bonneville faintly resembles Irving's representation of himself in *A Tour*" ("The Author as Professional," 248).

61. Irving, *Captain Bonneville*, 25.

62. Irving, *Captain Bonneville*, 25.

63. Irving, *Captain Bonneville*, 25.

64. Irving, *Captain Bonneville*, 25. Elsewhere Irving notes the westward migration of another species, bees: "The bees, according to popular assertion, are migrating like the settlers, to the West. An Indian trader, well experienced in the country, informs us that within ten years that he has passed in the Far West, the bee has advanced westward above a hundred miles" (*Captain Bonneville*, 18). More work can be done on Irving's interest in the domestication of bees as part of U.S. agri-expansion.

65. Irving, *Captain Bonneville*, 25.

66. Irving, *Captain Bonneville*, 25.

67. Irving, *Captain Bonneville*, 215.

68. Irving, *Captain Bonneville*, 210.
69. Irving, *Captain Bonneville*, 212.
70. Irving, *Captain Bonneville*, 212.
71. Irving, *Captain Bonneville*, 215.
72. Irving, *Captain Bonneville*, 214.
73. "Irving vacillated between a romantic and realistic view of the Indians, in much the same way that Cooper did," according to Littlefield; Irving presented a relatively "enlightened attitude toward the Indians, an attitude which formed the basis for ominous predictions, the truth of which the later works bear out" ("Washington Irving and the American Indian," 136).
74. Kendall Johnson argues, "Irving's ode to the imperial corporate form reflects both his concern over the power of American companies and implicit agreement with federal policies of Indian Removal" ("Caleb Cushing," 7). In addition, Kime observes that Irving's purchase of Sunnyside—his house in upstate New York—corresponded with the publication of his western pieces: "Home ownership and professional authorship were closely linked in his mind," adding, "Irving never in his career wrote so much, so well as when he produced his contributions to the literature of the American West" ("The Author as Professional," 238, 252).
75. Qtd. in Slotkin, *Regeneration through Violence*, 280.
76. As early as *The Sketch Book*, states Michael Schnell, Irving was engaged "with the new art of consuming food" ("The Tasteful Traveler," 111). For more on consumption in "The Legend of Sleepy Hollow," see Frederick Kaufman, "Gut Reaction."
77. B. Jones, *Washington Irving*, ix.
78. Dickens, *American Notes for General Circulation*, 130.
79. McGann, "Washington Irving," 349.

2. Civilizing Cattle

1. J. Cooper, *The Pioneers*, 229.
2. Susan Cooper followed her father, adding Fenimore to her own name, derived from the farm and her paternal grandmother's maiden name.
3. R. L. Johnson, *Passions for Nature*, 17.
4. See David Jones's introduction to Cooper's 1968 edition of *Rural Hours*, xxxvii–xxxviii.

5. Of the many names James Cooper gives the character—Natty Bumppo, Hawkeye, Deerslayer, and Pathfinder, among others—I use Leather-stocking in this chapter, as it was the ultimate name of the series itself and thus perhaps the most iconic of names for the man referred to only as "The Trapper" throughout *The Prairie*.

6. Turner writes that they descended from "a wealthy and aristocratic land speculator (William Cooper), who had been a gentleman promoter of backwoods settlement rather than a member of that kind of society" (*The United States*, 141). Virginia DeJohn Anderson observes: "By competing with local fauna, clearing away underbrush, and converting native grasses into marketable meat, imported animals assisted in the transformation of forests into farmland" (*Creatures of Empire*, 4).

7. Sweet, "Global Cooperstown," 541.

8. Wayne Franklin observes that the merino breed "was in great demand just then" because of "the Embargo Act's exclusion of European supplies," specifically of wool (*James Fenimore Cooper*, 182).

9. Nancy Shour writes that "this private, regional New York landscape is now a generically 'American' scene—just as it became for the River School painters" ("Heirs," 19). For more on Cooper family history see Franklin, *James Fenimore Cooper*.

10. J. Cooper, *The Pioneers*, 160.

11. J. Cooper, *The Pioneers*, 299.

12. Omi and Winant, *Racial Formation*, 3. Meanwhile, "American social and political policy towards Indians has been a two-hundred-year back-and-forth between assimilation and destruction," argues Deloria (*Playing Indian*, 5).

13. Myers, *Converging Stories*, 15.

14. Paul Jentz observes that in addition to showing his characters stereotypically as "noble savages," "his work also examines the complexities of racial identity and offers significant insight into the nuances of interracial relationships" (*Seven Myths*, 16). In addition, Timothy Sweet argues that reading the Coopers globally and ecocritically can be an exciting practice, "as an exploration of some possibilities and limitations of the global turn in ecocriticism, directed toward an object text that we have thus far regarded as especially local [*The Pioneers*]" ("Global Cooperstown," 541–42).

15. J. Cooper, *The Pioneers*, 160.

16. J. Cooper, *The Pioneers*, 161.

17. "Spurred by the accelerating industrialism of the nineteenth century," Antoine Traisnel writes, "hunting was gradually rendered obsolete by more efficient means of procuring animals, as in the introduction of factory farming and the industrialization of the slaughter, and by the taming of the nation's wild territories" ("American Entrapments," 43).

18. S. Cooper, *Rural Hours*, 94–95.

19. Shour writes of *The Pioneers*: "The plot of the novel is propelled by the conflicts of competing versions of the American pasts—the Indian past, the past of wilderness, the settlement past, the colonial past, the Revolutionary past—and the claims of the American present, as represented by the conflict of law and unethical exploitation of nature" ("Heirs," 20). Rochelle Johnson adds that to this historical reading, Susan Cooper adds the natural: "In fact, natural history—the formal study of species and natural phenomena—was never so popular as in the early decades of the nineteenth century" (*Passions for Nature*, 4).

20. According to Traisnel, "the frontier, in the American imaginary, is never simply a demarcation line separating the wild from the civilized but also the mythical soil in which the American dream is rooted." ("American Entrapments," 30).

21. J. Cooper, *The Prairie*, 97. The misuse of the term "buffalo" for American bison is not merely a modern issue. Dr. Battius corrects Leather-stocking at one point, in his typically pompous manner, "I am grieved when I find venator or hunter of your experience and observation, following the current of vulgar error. The animal you describe, is in truth a species of the bos ferus or bos sylvestris, as he has been happily called by the poets, but, though of close affinity it is altogether distinct, from the common Bubulus. Bison is the better word, and I would suggest the necessity of adopting it in future, when you shall have occasion to allude to the species" (*The Prairie*, 76–77). Cooper in theory leaves it up to the readers to determine which term is the more appropriate.

22. J. Cooper, *The Prairie*, 11. Lance Schachterle observes that in his introduction Cooper most clearly shows "the weirdness of the prairie" ("On *The Prairie*," 125).

23. J. Cooper, *The Prairie*, 90.

24. J. Cooper, *The Prairie*, 17, 11.

25. Matthew Sivils argues that in the novel, Leather-stocking "again witnesses the human blade of settlement cutting swaths of homesteads, farmland, towns, and outposts ever westward," becoming once more a kind of voiced conscience, that through him "Cooper champions a more holistic appreciation of the environment" ("Doctor Bat's Ass," 356). Also see H. N. Smith, Introduction, xii–xiii.

26. J. Cooper, *The Prairie*, 56.

27. J. Cooper, *The Prairie*, 56.

28. J. Cooper, *The Prairie*, 60.

29. Vance, "Man and Beast," 323.

30. *Cattle* began as a general term used to refer to any kind of marketable livestock—indeed, this was even a term for enslaved people during slavery, as a way to enact dehumanization against them. By the nineteenth century, though, it was being used primarily for the actual bovine.

31. Vance observes that "all humanity is seen to exist fundamentally in a predatory or exploitative relationship to the rest of the animal world" ("Man and Beast," 322).

32. J. Cooper, *The Prairie*, 198.

33. J. Cooper, *The Prairie*, 198.

34. J. Cooper, *The Prairie*, 201.

35. John L. Sutton argues that James Cooper helps to "establish the bison as an 'authentic' American image" ("Savory Bison Hump," 271–72). Vance adds: "Beyond the direct relations between men and other beasts, beyond the scientists' attempts to achieve a lucid zoology, *The Prairie* explores the question of man's own ambiguous identity as an animal with a difference" ("Man and Beast," 326).

36. H. N. Smith, Introduction, v.

37. J. Cooper, *The Prairie*, 44.

38. J. Cooper, *The Prairie*, 335.

39. Laura Mielke continues, "Natty and the Bush family have an uneasy relationship due to their differing conceptions of the prairie (natural space versus territory to be claimed)" (*Moving Encounters*, 44).

40. J. Cooper, *The Prairie*, 44.

41. Slotkin credits Cooper with transferring "the burden of the negative 'Indian' qualities from the red man to the common borderers and squatters—the vengeful, clannish Bush family in *The Prairie*, for

example" (*Regeneration through Violence*, 418). Meanwhile, Sarah Klotz reads the legal and cultural work being used to justify Euro-American expansion into obviously inhabited territory throughout the nineteenth century. "Since Native Americans were presumed to be an oral culture that had no writing systems and certainly no alphabet, they were viewed as intellectually behind Europeans, or closer to beasts than men," she argues, adding that "the right of discovery, which today seems a laughable justification for colonization, proceeds from cultural beliefs about literacy" ("The Red Man," 339).

42. It is fitting, then, that Leather-stocking's final resting place is "in a landscape that was, in 1827, just a little farther west than the farthest West that U.S. pioneers had attempted to settle," writes Stephanie LeMenager (*Manifest and Other Destinies*, 35). Marc Botha observes that "it is insecurity that defines the frontier experience of the settler-colonist, for whom surveyance and surveillance are regarded as twin imperatives in securing the property rights" ("Toward a Critical Poetics," 781). Moreover, Edward Larkin argues that in *The Prairie* Leather-stocking proves that he is "sympathetic to the culture of Native Americans, but ultimately allied to the interests of his fellow white men, who now have gained control of Native American lands" ("Time, Empire, and Nation." 221). For more on European readings of *The Prairie*, see Shields, "Savage and Scott-ish Masculinity"; and May, "The Romance of America."

43. R. L. Johnson and Patterson, introduction to *Rural Hours*, ix.

44. Buell, *The Environmental Imagination*, 221.

45. Sweet observes, "In the 1840s, with the frontier cycle over, eastern New York saw more out-migration than in-migration; many farmers moved west in hopes of gaining a freehold" ("Global Cooperstown," 557). Indeed, Lucy Maddox notes, "Cooper's father, James Fenimore Cooper, was instrumental in helping to establish the Agricultural Society of Otsego County as early as 1817, the first in New York" ("Susan Fenimore Cooper's Rustic Primer," 85). Against this agricultural intensification, she writes a natural component: "Cooper conceives of *Rural Hours* as her antidote, an attempt to make American readers appreciate what is left of the wilderness" (Irmschler, "Susan Fenimore Cooper's Ecology of Reading," 42).

46. S. Cooper, *Rural Hours*, 97.

47. Interestingly, the only other chores mentioned are also agricultural—Cooper mentions flocks as well as cattle; sheep and their wool are the next most significant element of her idea of appropriate agricultural domesticity, as with her father.

48. S. Cooper, *Rural Hours*, 100.

49. Kelly Clasen observes Cooper's focus on "rural surroundings—weather conditions, animal behaviors, agricultural practices, plant cycles, and human activity" ("Gender and Genre," 80). Richard Magee observes that Cooper is impressed by this dairy, "a prime model of domestic satisfaction," in "an extended narrative of community and rural values that is sustained longer than any other single episode in the book" ("Sentimental Ecology," 30).

50. S. Cooper, *Rural Hours*, 244–45.

51. S. Cooper, *Rural Hours*, 96.

52. Clasen argues, "*Rural Hours* promotes the adoption of an environmental ethos based on practical sustainability and conservation and rooted in her awareness of the interconnectedness of species" ("Gender and Genre," 90).

53. S. Cooper, *Rural Hours*, 108.

54. S. Cooper, *Rural Hours*, 108.

55. Allen, "The Sacred Hoop," 243.

56. Magee observes that Cooper "notes the Indians' degradation as her father does, but differs from him by noting that civilization is responsible for it" ("Sentimental Ecology," 34). He adds that she "acknowledges the displacement and alienation the Indians have a right to feel, as well as the debt that white Americans owe them. Significantly, this debt is not described in pecuniary terms, but in moral, specifically Christian, terms" ("Sentimental Ecology," 35).

57. S. Cooper, *Rural Hours*, 190.

58. Klotz writes, "In fact, the land had been under cultivation for ages, just not by Anglo-Americans" ("The Red Man," 358). Stephen Germic adds that these scenes "signify concerns about what, or who, bore the shock of the frontier violence attendant to Euro-American expansion and settlement" ("Land Claims, Natives, and Nativism," 480).

59. Clasen adds that Cooper "repeatedly references species decline to suggest the changes being wrought on her farming community by the

trend toward large-scale production and the corresponding deforestation" ("Gender and Genre," 89).

60. Josh Weinstein notes that her work is "predicated upon a Christian humility" ("Susan Cooper's Humble Ecology," 66). Tina Gianquitto adds a reading of this "natural theology" in Susan Cooper, in which "careful observation of the intricate structures and intimate connections of the natural world provides humans not only with a record of their accomplishments but with a model upon which to base their future actions" ("Noble Designs of Nature," 172). Nina Baym connects this decorousness to Susan Cooper's class association: "No Transcendental intuitionist, she believes in the biblical God of Christian Revelation" (*American Women of Letters*, 76).

61. S. Cooper, *Essays on Nature and Landscape*, 66.

62. R. L. Johnson and Patterson, introduction to *Susan Fenimore Cooper*, xxii.

63. S. Cooper, *Essays on Nature and Landscape*, 69.

64. S. Cooper, *Essays on Nature and Landscape*, 70. Barbed wire had been originally invented in 1867, before the publication of her essay; however, the more popular version of barbed wire, patented by Joseph Glidden, did not appear until 1874.

65. R. L. Johnson, "*Walden*, *Rural Hours*, and the Dilemma of Representation," 182.

66. Susan Cooper "believed that a failure to perceive nature's alteration would result in both an ill-conceived sense of American history and blind adherence to an ethic of 'progress,' which would inevitably damage the physical environment," according to Rochelle Johnson ("*Walden*, *Rural Hours*, and the Dilemma of Representation," 188).

3. Regional Cuisine and Cattle

1. Thoreau, *Journal XI*, 275.

2. Thoreau describes his dietary goals in the "Economy" chapter of *Walden*.

3. Thoreau writes: "I was pleased to read afterward, in Mourt's Relation of the landing of the Pilgrims in Provincetown Harbor, these words: 'We found great muscles (the old editor says that they were undoubtedly sea-clams) and very fat and full of sea-pearl; but we could not eat them, for they made us all sick that did eat, as well sailors as passengers,. . . . . but they were soon well again.' It brought me nearer to the Pilgrims to

be thus reminded by a similar experience that I was so like them" (*Cape Cod*, 74). Harding describes a conversation between Thoreau and an acquaintance, Jonathan Buffum, on the theme: "He was amused when Buffum warned him never to eat sea clams without first removing what he thought to be the clam's penis" (*Days of Henry Thoreau*, 396). Thoreau credits the advice of John Newcomb, the Welfleet Oysterman, as he is a native of the region and he possesses a deep knowledge of the foods of the area. Buffum, on the other hand, becomes a comic figure.

4. V. Anderson, *Creatures of Empire*, 75.

5. V. Anderson, *Creatures of Empire*, 99.

6. Even in colonial America, "English and other European settlers reenacted their predecessors' experience in seeking new territory on which to support proliferating herds" (V. Anderson, *Creatures of Empire*, 10).

7. Wagner, "Dining Out with Henry Thoreau," 2.

8. See Christie, *Thoreau as World Traveler*, 214.

9. Thoreau, *A Week*, 57.

10. Kenneth Allen Robinson gives a detailed description of Thoreau's focus on foods—specifically on wild foods: "Wherever he journeyed, Thoreau always sampled the indigenous foods and native fare" (*Thoreau and the Wild Appetite*, 84). Moreover, Thoreau's concerns with diet were largely due to his ongoing battle with tuberculosis; he believed that wild foods would benefit his health. See Leach, "Thoreau's Aphoristic Form."

11. Thoreau, *Walden*, 121–22.

12. Minott remained one of Thoreau's closest friends. For more on their relationship, see Harding, *Days of Henry Thoreau*, 455; Richardson, *Henry Thoreau*, 274–75; and Walls, *Henry David Thoreau*, 15. According to Percy Bidwell, New England farms were now "supplying the factory population with bread-stuffs, pork, beef, and wool." He also notes the other two areas of specialization were tobacco, in a "narrow strip of Connecticut River lowlands," and wool, in the hilly regions of western Massachusetts and Connecticut ("Agricultural Revolution in New England," 683–89).

13. Thoreau's concerns about the industrialized food economy were part of a larger cultural movement. Bill McKibben writes that "Americans came to full literary consciousness while much of their land was yet to be deforested, drained, cleared, developed. It is the fraught relationship between man and nature that suffuses many of the best American novels, poems,

and stories—and that many of the most eloquent and impassioned American essays take as their central subject" ("Nature and Environment").

14. Walls, *Henry David Thoreau*, xi.
15. Barthes, "Toward a Psychosociology," 29.
16. Bakhtin, *Rabelais and His World*, 285.
17. Denise Gigante analyzes "the metaphor of consumption in the field of representation" in the work of British romanticists in a way that applies to Thoreau's writing as well (*Taste*, 2). For a discussion of Thoreau in food politics see Fiskio, "Sauntering across the Border." Moreover, the field of animal studies—as seen, for example, in Donna Haraway's *The Companion Species Manifesto* and *When Species Meet*—has blossomed in the twenty-first century, and my interest in food studies intersects in relevant ways in terms of Thoreau's interactions with animals (collecting and consuming) in his excursion narratives.
18. Thoreau, *A Week*, 25.
19. Thoreau, *A Week*, 26.
20. McPhee, *The Founding Fish*, 219. He names Thoreau the most "passionate" defender of the shad, calling him "radically and almost subversively anthropomorphic in his regard for the founding fish." McPhee quotes an 1839 Massachusetts Zoological and Botanical Survey: "The Concord shad have almost entirely disappeared, their ascent having been cut off by dams" (219–20).
21. Thoreau, *A Week*, 33.
22. Thoreau, *A Week*, 37.
23. Buell discusses Thoreau's early conservationist company briefly in "Thoreau and the Natural Environment" (173–74).
24. See Neely, "Embodied Politics"; and Dolan, "Diet and Vegetarianism."
25. Thoreau, *Walden*, 210.
26. Linck Johnson considers the "intensity of Thoreau's apostrophe to the shad" as one of Thoreau's connections to the death of his brother and the fragility of life (*Thoreau's Complex Weave*, 56). Philip Cafaro discusses bream in terms of sacrificing representative animals to scientific study in Thoreau's work—and this applies as well to the moose: "We cannot map out food webs, for example, without censusing the plants present in an area and determining what the animals are eating—sometimes killing them in the process" ("Thoreau on Science and System").
27. Thoreau, *The Maine Woods*, 55.

28. Thoreau, *The Maine Woods*, 57.
29. Thoreau, *The Maine Woods*, 57.
30. Thoreau, *The Maine Woods*, 38.
31. Thoreau, *The Maine Woods*, 16.
32. Thoreau, *The Maine Woods*, 24.
33. Thoreau, *The Maine Woods*, 129.
34. Thoreau, *The Maine Woods*, 58.
35. Thoreau spells the name "Aitteon," but the Penobscot spelling is "Attean." This chapter will use the Penobscot spelling except for in direct quotations. Thatcher was "a Bangor lumber merchant" (Kucich, "Lost in the Maine Woods," 28).
36. Van Ballenberghe continues: "Square bodies conserve heat in climates where midwinter low temperatures reach sixty degrees below zero. Long noses enhance aquatic feeding and extend the reach of moose to feed on tall shrubs" (*In the Company of Moose*, 3).
37. Thoreau, *The Maine Woods*, 81. For further discussion on Thoreau and Native American peoples, see Rose, "Tracking the Moccasin Print"; and Sayre, *Thoreau and the American Indians*.
38. Williamson, *History of the State of Maine*, 135.
39. Thoreau, *The Maine Woods*, 110, 115.
40. Thoreau, *The Maine Woods*, 122.
41. Thoreau, *The Maine Woods*, 97; Thoreau, *Journal, Vol. 7*, 52.
42. Richardson, *Henry Thoreau*, 302; Gura, "Thoreau's Maine Woods Indians," 375. Kucich adds: "Thoreau's view of the moose has moved from tourist trophy to full personhood" in the course of his description of the hunting and butchering scene ("Lost in the Maine Woods," 30).
43. Thoreau, *The Maine Woods*, 117.
44. Thoreau, *The Maine Woods*, 132.
45. Thoreau, *The Maine Woods*, 132.
46. Thoreau, *The Maine Woods*, 129.
47. Thoreau, *The Maine Woods*, 140–41.
48. Thoreau, *The Maine Woods*, 147. Compare this depiction with Thoreau's description of meeting Louis Neptune—who is hunting musquash for future expeditions—in "Ktaadn." In the earlier scene, Thoreau theorizes degradation in a way that is at once patronizing—even offensive—yet also recognizes that in a larger sense, "In the progress of degradation, the distinction of races is soon lost" (78).

49. Thoreau, *The Maine Woods*, 123.
50. Thoreau, *The Maine Woods*, 304.
51. Thoreau, *The Maine Woods*, 153.
52. Thoreau, *The Maine Woods*, 276.
53. Thoreau, *The Maine Woods*, 288.
54. Thoreau, *The Maine Woods*, 283.
55. Thoreau, *The Maine Woods*, 288.
56. Thoreau, *The Maine Woods*, 269.
57. Thoreau, *The Maine Woods*, 233.
58. Cows would not have to be entirely excluded from such an alternative paradigm, however. In Sara Orne Jewett's story "A White Heron," for example, set in the Maine woods of the 1880s, Sylvia's grandmother keeps a single dairy cow that seems to pretty much forage for itself in some rough pasturage. This would surely not be at odds with a Thoreauvian localist model.
59. James Finley discusses the essay in terms of "the poetic action of turning raw material into text" (*"Who* Are We?," 349).
60. Harding, *Days of Henry Thoreau*, 466.
61. Thoreau, *The Maine Woods*, 119.
62. The exact number of casualties has been a subject of controversy; I get this number from Ryan Schneider, "Drowning the Irish," 467.
63. Thoreau, *Cape Cod*, 147. His sublime understanding following the tragedy parallels in some ways his near panic while climbing Mount Katahdin in *The Maine Woods. Cape Cod* alludes to John Thoreau's death in 1842 through the name of the ship, the *St. John*. For a discussion on the shipwreck section of *Cape Cod*, see Morgan, "Requiem for the *St. John*," 23–34.
64. The first four chapters were published in *Putnam's* in 1855—including the shipwreck scene and Thoreau's clam-eating adventure. However, *Putnam's* rejected the "Welfleet Oysterman" chapter and cut off publication at that point (Moldenhauer, "Historical Introduction," 267–75).
65. Katie Simon observes that *Cape Cod* potentially thwarts "a more traditional environmental preservationist message," adding that "this failure may also enable a different message now, during a moment of global economic and environmental crisis" ("Affect and Cruelty," 246).
66. Kurlansky, *The Big Oyster*, 16.
67. Thoreau, *Cape Cod*, 66.

68. Thoreau, *Cape Cod*, 45.
69. Thoreau gives a detailed description of the various "Mollusca" of the region (*Cape Cod*, 85–86).
70. Thoreau, *Cape Cod*, 56.
71. Thoreau, *Cape Cod*, 56–57.
72. Thoreau, *Cape Cod*, 67.
73. Thoreau, *Cape Cod*, 73–74.
74. Thoreau, *Cape Cod*, 67.
75. Thoreau, *Cape Cod*, 147.
76. Ronald Morrison writes: "Implicit in *Cape Cod* is Aldo Leopold's view that cultures are reflections of 'the wilds that gave them birth,' that the particular wilderness out of which a culture grows is the starting point for understanding its cultural heritage" ("Time, Place, and Culture," 218).
77. Thoreau, *Cape Cod*, 130.
78. Thoreau, *Cape Cod*, 158.
79. Thoreau, *Cape Cod*, 164.
80. Thoreau discusses explorers and their livestock in some of his quotations in *Cape Cod*. The Spanish, who had been on "Isola della Rena" around 1556–65, left cattle behind them for the French and English explorers to find. "Charlevoix says they ate up the cattle and then lived on fish. Haliburton speaks of cattle left there as a rumor" (*Cape Cod*, 190).
81. Thoreau, *Cape Cod*, 164.
82. Another kind of grass does not appeal to Thoreau. The kelp that is a constant sight in Cape Cod only appeals to Thoreau in extremity: "This species looked almost edible, at least, I thought that if I were starving I would try it. One sailor told me that the cows ate it" (*Cape Cod*, 53).
83. Thoreau, *Cape Cod*, 169.
84. Thoreau, *Cape Cod*, 169.
85. Thoreau, *Cape Cod*, 33.
86. Walls, *Henry David Thoreau*, 280.
87. Gura, "A Wild, Rank Place," 144.
88. Sharon Talley discusses the diet of the Newcomb family, "an unconventional mixture of the wild and the cultivated" that "is largely dependent on the motley harvest they are able to reap from the ocean and their garden" ("Thoreau's Taste," 88). This motley combination of wild and civilized ultimately appeals to Thoreau.

89. Pinsky, Introduction, ix. Pinsky further states that Thoreau's version of travel narrative "is an enterprise brilliantly pursued a generation later by the travel books of Mark Twain" (xviii).

90. Michael Breitweiser writes, "*Cape Cod* is a book in which many sorrowful things are described but in which there is little sorrow" (*National Melancholy*, 147). In their introduction to *Thoreauvian Modernities* (2013), François Specq and Laura Dassow Walls argue that *Cape Cod* "offers a more complex and 'pluralized' understanding that points toward the openness of postmodern thinking—or, more accurately 'nonmodern' thinking—understood not as an ironical game depriving the world of any real significance but as an enlargement of the literary, social, and natural fields across four fundamental temporal and spatial dimensions" (5).

91. Thoreau, *Cape Cod*, 144.

92. Pinsky, "Comedy, Cruelty, and Tourism," 84.

93. See especially Walls, *Henry David Thoreau*, 404–44.

94. Thoreau, *A Week*, 225.

4. Cattle and Sovereignty

1. Reprinted in Winnemucca Hopkins, *Newspaper Warrior* (2015). I use the twenty-first-century spelling, "Paiutes," throughout. Writing in the nineteenth century and in her second language, Winnemucca's spelling varied; "Pah-Utes" and "Piutes" are examples of this.

2. For more information of the history of the Bannock War of 1878, see Heaton, *The Shoshone-Bannocks*, 48–51.

3. Fiskio, "Where Food Grows on Water," 242.

4. United Nations, "United Nations Declaration on the Rights of Indigenous Peoples," 22–23.

5. Nabhan, *Cultures of Habitat*, 258. He adds: "Nancy Turner reports that the Salish people of the coastal rain forests of Washington have used banana slugs as a poultice for cuts and wounds because these slugs consume a certain set of plants that collectively have medicinal value. But there are fewer than ten Salish speakers left in this world" (258).

6. Peter Iverson observes that "Indian farmers lost their lands to the growing Spanish enterprise" of imperialistic agriculture (*When Indians Became Cowboys*, 8).

7. See Knowlton, *Cattle Kingdom*, xviii–xx. Iverson argues that "cattle ranching emerged, therefore, as a symbol for a new day. But that new day was

long in coming, following an extended period of non-Indian expansion, Indian resistance, and the drawing of reservation lines" (*When Indians Became Cowboys*, 14). Virginia DeJohn Anderson adds: "If the arrival of European diseases had the greatest impact on native societies, the introduction of Old World animals proved at least as wide-ranging in its effects" (*Creatures of Empire*, 184).

8. Myers, *Converging Stories*, 5.

9. Outka, *Race and Nature*, 7.

10. Karl Jacoby declares that "Americans have often pursued environmental quality at the expense of social justice" (*Crimes against Nature*, 198). He adds: "Of all the decisions any society must make, perhaps the most fundamental ones concern the natural world, for it is upon the earth's biota—its plants, animals, waters, and other living substances—that all human existence ultimately depends" (1). In addition, "Much as the conquest of the West reshaped ideas about wilderness, it also led to the creation of an extensive reservation system," writes Mark David Spence, providing a famous example: "Black Elk understood all too well that wilderness preservation went hand in hand with native dispossession" (*Dispossessing the Wilderness*, 3–4).

11. Winnemucca Hopkins, *Newspaper Warrior*, 139.

12. For example, "And the seven years of plenteousness, that was in the land of Egypt, were ended" (Genesis 41:53, King James version). Winnemucca combined her traditional faith with Christianity and would have known familiar biblical verses. Her allusion, therefore, to the end of a time of plenty would be not only readily recognizable by her audience but would also be rhetorically powerful for her argument.

13. Lewis and Clark, *Journals*, 260; bracketed material in source.

14. Zanjani, *Sarah Winnemucca*, 8. For more discussion of the history of the Northern Paiutes, see Rifkin "Finding Voice in Changing Times," 146–52. The Paiutes of Nevada lived where the current "Nevada Test Site" region has been located in contemporary times. In an analysis of late-twentieth-century cultural politics, T. V. Reed notes that it is the Native Americans of this region who historically "have suffered most directly the effects that the patriarchal military-corporate-scientific complex have inflicted on this particular landscape" (*The Art of Protest*, 236).

15. See Ruoff, "Early Native American Women Authors," 85–86. Indeed, Winnemucca's real life was dramatic enough to justify memorialization.

In addition to her dramatic horseback ride during the Bannock War, she also developed a school for Native American children that included traditional knowledge and language with English—a revolutionary concept during her time (Carpenter and Sorisio, Introduction, 4–9).

16. Winnemucca could ride exceedingly well, and her family was sometimes paid in horses. Paiutes were sometimes labeled "horse thieves" or "cattle thieves" interchangeably by ranchers seeking conflict. Gae Whitney Canfield notes that "horses would greatly increase the food-gathering capabilities of the bands" (*Sarah Winnemucca*, 6). Cattle, though, remains the more rhetorically powerful symbol for Winnemucca.

17. Winnemucca Hopkins, *Newspaper Warrior*, 139.

18. Karen Kilcup refers to the Bannock War as an example of "resource wars" (*Fallen Forests*, 210–11). Michelle Kohler finds in Winnemucca's writing "a profound ambivalence toward writing" itself, studying scenes in *Life among the Piutes* of the official letters or "rag friends" of Truckee, and their subsequent betrayals throughout the autobiography that are part of the paradox of "national rhetoric and the literal removal of Indians" ("Sending Word," 54, 73). Heidi Hanrahan adds that the relationship between Winnemucca and the Peabody sisters "emerges as one of mutual influence, construction, and cooperation, and provides readers with a more complicated understanding of the dynamic, dialogic interactions between Native American authors and their white editors" ("'[w]orthy the imitation of the whites,'" 120). Finally, Noreen Grover Lape also discusses Winnemucca's liminality in navigating multiple and opposing cultures ("I Would Rather Be with My People").

19. Powell, "Princess Sarah, the Civilized Indian," 64.

20. Sorisio, "Playing the Indian Princess?," 2. In addition, Sorisio argues that Powell conflates "civilized Indian" and "Indian princess": "The Indian princess shares with the exemplary Indian a capacity for civilization . . . whereas Indian princesses were to conform to the audiences' desires for authentic dress" (17). Winnemucca manages to exist in both spaces during her performances, thereby "disrupt[ing] the manifest manners not only of her time but also of scholarship today" (4).

21. Quoted in Jackson, *Century of Dishonor*, 395–96.

22. Fiskio, "Where Food Grows on Water," 238–39.

23. Joseph F. Glidden patented modern barbed wire in 1874; it quickly became a technological revolution in cattle ranching in the West (Krell, *The Devil's Rope*, 24).

24. "Whites' livestock consumed the grasses the Paiutes depended on for seed food," writes Sally Zanjani (*Sarah Winnemucca*, 17). In addition, according to Myers, "settlers plowed under or otherwise converted to agriculture over 600 million acres of grassland, including 100 million acres of American Indian lands forfeited under the Dawes and Curtis Acts," adding, "Bison on those lands were reduced from around 30 million to 500 individuals" (*Converging Stories*, 3).

25. Knowlton, *Cattle Kingdom*, 218. David Igler adds: "Miller specifically advised the foreman against hiring Paiute Indians from the nearby Malheur River Reservation, since tensions ran high between eastern Oregon settlers and Indians" (*Industrial Cowboys*, 152).

26. Winnemucca Hopkins, *Life among the Piutes*, 86.

27. Winnemucca Hopkins, *Life among the Piutes*, 81.

28. Winnemucca Hopkins, *Newspaper Warrior*, 223.

29. Winnemucca Hopkins, *Newspaper Warrior*, 225.

30. Winnemucca Hopkins, *Newspaper Warrior*, 224. For more on Paiutes and agriculture, see Canfield, *Sarah Winnemucca*, 10–11.

31. Winnemucca Hopkins, *Newspaper Warrior*, 223. Joni Adamson discusses the genre of environmental justice, which "brings the voices of writers and artists into the conversation as it explores not only politics and policy, but poetics and pedagogy" ("Medicine Food," 213).

32. Sorisio, "Playing the Indian Princess?," 3–6. Joanna Cohen Scherer writes about Winnemucca's portraiture during her career. Noting that her 1880 DC trip photo is one of the few that does *not* show her in "Indian Princess costume," she adds: "Few Indian women appear in formal delegation photographs, and her inclusion underscores her important status in her family and her position as a spokesperson for her tribe" ("Public Faces of Sarah Winnemucca," 184).

33. Sarah Winnemucca Hopkins, "Statement of Mrs. Hopkins," 11, House Subcommittee on Indian Affairs, 1884, Sally Zanjani Papers, 2013-4, Special Collections, University Libraries, University of Nevada, Reno.

34. Winnemucca Hopkins, "Statement of Mrs. Hopkins," 18.

35. Cari Carpenter notes that "since Sarah Winnemucca's grandfather, known as Truckee, had welcomed white settlers to the area, the fam-

ily had enjoyed a certain political power in relation to white society" ("Sarah Winnemucca Goes to Washington," 87). For more about controlled spaces in Winnemucca's writing, see Senier, *Voices of American Indian Assimilation and Resistance*.

36. Rosalyn Collings Eves writes that "those along the frontier (like Winnemucca) were viewed as foreign and therefore subject to imperial conquest" and moved to reservations that represented Foucault's "disciplinary space" ("Finding Place to Speak," 1–2). "Because livestock tend to be discussed in terms of the ecological alterations they produced," Virginia DeJohn Anderson notes, "the effects of their presence on people are largely indirect, mediated by the environment itself" (*Creatures of Empire*, 5). Also see Spence, *Dispossessing the Wilderness*, 60–65.

37. Greg Garrard notes that although "the history of the colonisation of America has to be seen, at least in large part, in ecological terms," modern readers should avoid the stereotype of the "Ecological Indian" (*Ecocriticism*, 133). Fiskio concludes: "Indigenous North American literature describes the impact of these colonial tactics and the ongoing resistance to settler colonialism through the practice of indigenous subsistence and foodways and efforts to maintain and recover indigenous food sovereignty" ("Where Food Grows on Water," 238–39). For more on ecocriticism see Pellow and Brulle, "Power, Justice, and the Environment," 1–21.

38. Winnemucca Hopkins, *Life among the Piutes*, 209.

39. As Cary Wolfe notes, "If the frame is about rules and laws . . . then to live under biopolitics is to live in a situation in which we are all always already (potential) 'animals' before the law—not just nonhuman animals according to zoological classification, but any group of living beings that is so framed," going on to give examples of slavery and colonialism, to which I add Indian removals (*Before the Law*, 10).

40. Winnemucca Hopkins, "Statement of Mrs. Hopkins," 5.

41. Winnemucca Hopkins, "Statement of Mrs. Hopkins," 10–11.

42. Winnemucca Hopkins, *Newspaper Warrior*, 203.

43. Winnemucca Hopkins, *Newspaper Warrior*, 223.

44. Winnemucca describes how Natchez came by his name: "He would say, pointing to my brother, 'my Natchez'; he always said this. So the white people called one of my brothers Natchez, and he has had that name to this day" (*Life among the Piutes*, 33).

45. Winnemucca Hopkins, *Life among the Piutes*, 219.

46. Ruoff, "Early Native American Women Authors," 85.
47. Howe, "Tribalography," 118.
48. Sneider, "Gender, Literacy, and Sovereignty," 264. See also Ruoff, "Early Native American Women Authors," 87; and Canfield, *Sarah Winnemucca*, 86–90.
49. Quoted in Kohler, "Sending Word," 52–53.
50. Katharine Rodier describes the influence of the Peabody sisters on Winnemucca as mutually beneficial, as the sisters combined "their preferred styles of intercession, both indirect and overt, with Winnemucca's own outspoken public aims in what emerges as a group portrait of female agency on the Paiute behalf" ("Authorizing Sarah Winnemucca," 108–9). Cheryl Walker notes that "both Sarah and her brother Natchez try to engage white sympathy by means of affecting transpositional rhetoric emphasizing parallels between Indians and whites" (*Indian Nation*, 153).
51. Zanjani, *Sarah Winnemucca*, 1–2.
52. Powell quoted in Carpenter, "Sarah Winnemucca Goes to Washington," 96. The term was coined by Gerald Vizenor, who adds: "Native survivance is an active sense of presence over absence, deracination, and oblivion; survivance is the continuation of stories, not a mere reaction, however pertinent" ("Aesthetics of Survivance," 1).
53. Ernest Stromberg adds: "While 'survival' conjures images of a stark minimalist clinging at the edge of existence, survivance goes beyond mere survival to acknowledge the dynamic and creative nature of Indigenous rhetoric" ("Rhetoric and American Indians," 1).
54. Zanjani, *Sarah Winnemucca*, 1. Adamson describes the First Indigenous Peoples' Global Consultation on the Right to Food and Food Sovereignty as exemplifying "the persistent responses of indigenous North American communities to broken treaties between First Nations and the expansionist U.S." ("Medicine Food," 215). David L. Moore comments that "in spite of her own and her people's losses, her narrative holds the door open to American community" (*That Dream Shall Have a Name*, 162).
55. See Canfield, *Sarah Winnemucca*, 191; and Zanjani, *Sarah Winnemucca*, 107.
56. Hittman, *Wovoka and the Ghost Dance*, 319.
57. Omi and Winant, *Racial Formation*, 80.
58. Louis S. Warren writes that the Paiutes' "growing reliance on cash made them ever more dependent on wages as the century wore on and rapid environmental changes continued. . . . By 1890, workplace

demands kept Paiutes on the ranches where they worked for all but two or three weeks of the year" (*God's Red Son*, 88). Warren places Wovoka and Winnemucca in conversation: "These Paiute testimonies convey a powerful sense of something broken that cannot be put back, of an old world shattered, of a people with no hope but to live in a strange new world and anticipate a new age ahead. From just such sensibilities was the new religion [the Ghost Dance] born" (102). In addition, Iverson notes that the Dawes Act, the General Allotment Act of 1887, "parceled out 160-acre, quarter-section plots to individual Indians. Indians could not boycott these proceedings. Government agents were empowered to make the selection for Indians who chose initially not to take advantage of this arrangement" (*When Indians Became Cowboys*, 29).

59. "Sarah Winnemucca," in Winnemucca Hopkins, *Newspaper Warrior*, 292.
60. Ruoff, "Early Native American Women Authors," 86.

5. Cowboys *Are* Indians

1. According to Donna Bradley, Sebastian Vizcaino named the Kumeyaay lands "San Diego" in 1602 (*Native Americans of San Diego County*, 7). I use the term "Kumeyaay" except in quotations and references throughout the chapter. For more on Kumeyaay history in California, see http://www.kumeyaay.info/kumeyaay/.
2. For a discussion of the legal rhetoric of *The Squatter and the Don*, see Ramirez, "American Imperialism in the Age of Contract"; Alemán, "Novelizing National Discourses"; and de la Luz Montes, "María Amparo Ruiz de Burton." In addition, John-Michael Rivera argues that *The Squatter and the Don* is "the first work of fiction to address the treaty" of Guadalupe Hidalgo directly (*The Emergence of Mexican America*, 98).
3. Carey McWilliams notes that "the social structure of Spanish California resembled that of the Deep South: the *gente de razón* were the plantation owners; the Indians were the slaves; and the Mexicans were the California equivalent of 'poor white trash'" (*North from Mexico*, 64). For more discussion of cattle, agriculture, and the economy following the Treaty of Guadalupe Hidalgo, see *North from Mexico*, 64–66.
4. Ruiz de Burton, *The Squatter and the Don*, 32.
5. According to Pablo Ramirez, the Land Act of 1851 "established a commission to investigate the validity of Mexican Americans' once-inalienable land titles," though this was largely ineffective and was used as a tool

for U.S. squatters ("American Imperialism in the Age of Contract," 431). Jesse Alemán notes that the No Fence Law "not only called into question Spanish-Mexican land grants but legalized the romantic notion of Manifest Destiny in general" ("Novelizing National Discourses," 41). Also see V. Anderson, *Creatures of Empire*, 77–78.

6. Ruiz de Burton, *The Squatter and the Don*, 155.
7. Saldívar, "Nuestra América's Borders," 162.
8. For a discussion of survivance, see Powell, "Rhetorics of Survivance"; and Vizenor, *Survivance*.
9. Lisbeth Haas notes: "She argues, for example, that the plight of Californio rancheros was aggravated by state laws that enabled settler-farmers to preempt land while the title was being confirmed" (*Conquests and Historical Identities*, 78).
10. See, for example, Alemán, "Citizenship Rights and Colonial Whites."
11. Corinna Barrett Percy writes that "she suppresses and stereotypes the Indians, who are also unjustly treated" ("Reformers, Racism, and Patriarchy," 116). Priscilla Ybarra observes that "California state laws quickly stepped in to support an imperial agenda that discouraged Anglo-Americans from working with Mexican Americans" (*Writing the Goodlife*, 43). Also, while Paul Outka largely discusses African American writings, I apply him here to a broader, diverse racialized community. Stacy Alaimo reads in the tendency to moralize "nature," "the erection of culture above a denigrated nature has fortified racist ideologies by charting a chain of being in which African, Mexican, and Native Americans dwell 'closest' to a debased nature" (*Undomesticated Ground*, 19).
12. Szeghi, "The Vanishing Mexicana/o," 105.
13. According to González, "Before the U.S. conquest in 1848, Californio rancheros had often kidnapped Indians from rancherías and forced them to work as ranch hands" (*Chicano Novels*, 89–90). Richard Slatta adds: "Spanish terminology, equipment, and technique spread from Texas and California throughout the western United States. In the 1880s a ranch foreman was called 'major domo' (from the Spanish term *mayordomo*), whether he was in eastern Oregon, southern Idaho, or Nevada. Cowboys in the region were called vaqueros" (*Cowboys of the Americas*, 7).
14. Ruiz de Burton, *The Squatter and the Don*, 44.
15. Ruiz de Burton, *The Squatter and the Don*, 44.
16. Ruiz de Burton, *The Squatter and the Don*, 24.

17. Ruiz de Burton, *The Squatter and the Don*, 46.
18. Krell, *The Devil's Rope*, 24.
19. Ruiz de Burton, *The Squatter and the Don*, 50.
20. Melanie V. Dawson observes that Ruiz de Burton's squatter characters "destabilize an agrarian economy and its people, wreaking emotional havoc" ("Ruiz de Burton's Emotional Landscape," 44). For further discussion of wheat in Norris's work, see Dolan, *Beyond the Fruited Plain*, 176–209.
21. Pérez, *Remembering the Hacienda*, 128. David Luis-Brown adds that in 1856 Captain Burton "led his troops in a confrontation with the Indians portrayed in Jackson's novel [*Ramona* (1884)]—the Luiseño Indians of Temecula, in San Diego County" ("'White Slaves' and the 'Arrogant Mestiza,'" 823).
22. Gilbert, "Articulating Identity," 19.
23. For more on Ruiz de Burton's friendship with Mariano Vallejo, see Warford, "An Eloquent and Impassioned Plea"; and Alemán, "Citizenship Rights and Colonial Whites."
24. Ruiz de Burton, *The Squatter and the Don*, 144.
25. Pérez, *Remembering the Hacienda*, 111–12.
26. Alemán, "Citizenship Rights and Colonial Whites," 17, 22. In all this history, Szeghi adds, "it is striking that she [Ruiz de Burton] never acknowledges the simple fact that American Indians are California's original inhabitants" ("The Vanishing Mexicana/o," 102).
27. Saldívar, "Nuestra América's Borders," 162.
28. Karen Kilcup observes that the novel's "cattle herding connects with economics rather than spirituality or pleasurable retreat." ("Writing against Wilderness," 373).
29. Ruiz de Burton, *The Squatter and the Don*, 228.
30. Ruiz de Burton, *The Squatter and the Don*, 228–29.
31. Ruiz de Burton, *The Squatter and the Don*, 229.
32. Ruiz de Burton, *The Squatter and the Don*, 229; my translation.
33. There have been many discussions of Ruiz de Burton's complicated relationship with race and ethnicity in both of her novels. See Luis-Brown, "'White Slaves' and the 'Arrogant Mestiza'"; Aranda, "Contradictory Impulses"; and Thomas, *The Literature of Reconstruction*.
34. Ruiz de Burton, *The Squatter and the Don*, 230.
35. Szeghi, "The Vanishing Mexicana/o," 107.

36. Szeghi, "The Vanishing Mexicana/o," 106.
37. Ruiz de Burton, *The Squatter and the Don*, 231.
38. Ruiz de Burton, *The Squatter and the Don*, 230.
39. For Martí, Gillman writes, this regional connection is about "bring[ing] together the two oppressed groups for which they speak." Martí's Stowe-Jackson composite author "becomes an inter-ethnic, international figure capable of speaking to both the limits and possibilities of the multiple racial and national aspirations of Latin America and the Caribbean" ("The Squatter," 141–42).
40. Kilcup, "Writing against Wilderness," 361. Amelia M. de la Luz Montes agrees: "Sentimentalism is a mode that has designs upon the world" ("*Es Necesario Mirar Bien*," 26).
41. Luis-Brown, "'White Slaves' and the 'Arrogant Mestiza,'" 814. Elisa Warford also places the novel among "different genres that are well suited to social protest, including verbatim legislation, the jeremiad, sentimental romance, and naturalism" ("An Eloquent and Impassioned Plea," 7).
42. In his discussion, Pérez notes: "Agricultural and ranching products function metaphorically in this scene, with hacienda products a figuration for Californio ranch community and grain or (commercial) crops a metonym for the threat posed by the squatters" (*Remembering the Hacienda*, 70).
43. Ruiz de Burton, *The Squatter and the Don*, 15.
44. Ybarra notes the double meanings of Mercedes: "mercy" and "land grants" (*Writing the Goodlife*, 40).
45. Ruiz de Burton, *The Squatter and the Don*, 215.
46. Ruiz de Burton, *The Squatter and the Don*, 292–93.
47. Melanie Dawson writes: "It insists on a regional and historical narrative where details about feeling—not land—form the basis of its verisimilitude" ("Ruiz de Burton's Emotional Landscape," 48). "The book's most detailed descriptions of cattle herding only gesture toward pastoral," adds Kilcup, "for they describe a drive under duress to Clarence's property" ("Writing against Wilderness," 374).
48. Ruiz de Burton, *The Squatter and the Don*, 290.
49. Ruiz de Burton, *The Squatter and the Don*, 291. The majordomo may also have been a Mexican character.
50. Ruiz de Burton, *The Squatter and the Don*, 293.
51. Ruiz de Burton, *The Squatter and the Don*, 292–94.

52. Ruiz de Burton, *The Squatter and the Don*, 18.
53. Jennifer Tuttle focuses on health and bodies in *The Squatter and the Don*: "In her attempt to grant voice and agency to the conquered Californios, María Amparo Ruiz de Burton deftly invokes health tourism and the promotion of Anglo American migration to California, using the discourse against itself" ("The Symptoms of Conquest," 61).
54. Thomas, "Ruiz de Burton, Railroads, Reconstruction," 875.
55. Ruiz de Burton, *The Squatter and the Don*, 309.
56. Ruiz de Burton, *The Squatter and the Don*, 309.
57. Slatta, *Cowboys of the Americas*, 3.
58. Stanford's own racial biases are on display in this scene. Brook Thomas observes that the railroad monopolists like Stanford and Collis Huntington were "much more intent on turning freedmen, Native Americans, and Chinese immigrants into productive workers than in making them equal political citizens" ("Ruiz de Burton, Railroads, Reconstruction," 881).
59. For more on Thoreau's reading of cattle and U.S. expansion, see chapter 3.
60. Szeghi, "The Vanishing Mexicana/o," 109.
61. Marissa López writes, "We see the human body, the material world, capital, the natural world, and social institutions like companies and nations all taking shape and expanding simultaneously" over the course of the nineteenth century ("Feeling Mexican," 188). Also see Ybarra, *Writing the Goodlife*, 136.
62. Ruiz de Burton, *The Squatter and the Don*, 265.
63. Chvany, "Those Indians," 111. González adds that "Chapo, an Indian servant at the Alamar rancho, singlehandedly derails the modernist project of national allegory as completely as any monopoly corporation" (*Chicano Novels*, 104).
64. Chvany, "Those Indians," 112.
65. Rodrigo Lazo writes that Ruiz de Burton "is not always successful" in her attempts to keep her indigenous characters peripheral to the action in her writings, implying that she intends to do just that ("Trans-American Literature," 10). Chvany observes that it is noteworthy that a novel about railroads, written by an author often accused of racism, never mentions Chinese people; he goes on to analyze the character Tisha, the novel's only (and very peripheral) African American character ("Those Indians," 110).
66. Ruiz de Burton, *The Squatter and the Don*, 143.

67. Ruiz de Burton, *The Squatter and the Don*, 203.
68. The one other mention of Kumeyaay is of one helping the broken Victorniano after his frostbite and paralysis following the snowstorm. Here, not only is he called the "Indian boy," but he is also called a "servant boy," perhaps giving him a more domestic appearance than the cattle ranchers/drivers of earlier in the novel—back when they owned cattle.
69. Ruiz de Burton, *The Squatter and the Don*, 327.
70. Ruiz de Burton, *The Squatter and the Don*, 327.
71. Contreras, "I'll Publish Your Cowardice," 219.
72. Alemán observes that the marriage itself is not truly a romantic affair, in spite of itself, "less a culmination of their romance and more a pragmatic response to the family's historical conditions" ("Novelizing National Discourse," 42). In addition, Margaret Jacobs observes that, as opposed to the English and U.S. laws, "under Spanish colonial and Mexican law, women retained their property rights after marriage and did not fall under coverture" ("Mixed-Bloods, Mestizas, and Pintos," 225).
73. Sánchez and Pita, Introduction, 7–8.
74. Sánchez and Pita, *Conflicts of Interest*, 539.
75. Castillo, Introduction, xiii. Sánchez and Pita add: "Her writing, in effect, becomes a space for movement between various social and national discourses and for representation of this outsider/insider dichotomy, serving not only as a site for construction of the collectivity, but as a forum for voicing the concerns of 'la raza'" (*Conflicts of Interest*, 546).
76. More work can be done on Ruiz de Burton's rhetoric in the novel— literary and philosophical works as well as legal briefs are mentioned throughout. González observes: "The narrator and three of the novel's main characters quote Herbert Spencer, Thomas Carlyle, Charles Dickens, Ralph Waldo Emerson, Joseph Addison, Charles Fourier, and William Ellery Channing, as if they were household names" (*Chicano Novels*, 68).
77. Ybarra, *Writing the Goodlife*, 58.
78. Alemán, "Novelizing National Discourse," 38.

6. Southern Cuisine without Cattle

1. I use Robert B. Stepto and Jennifer Rae Greeson's edition of *The Conjure Stories*, which presents the fourteen conjure stories chronologically.

2. Chesnutt, "Mars Jeems's Nightmare," 98. Throughout, I transcribe Chesnutt's quotes into standardized spelling, though with grammatical and stylistic flourishes intact, in parentheses for ease of reading.
3. I use the common term "pigs" throughout this chapter as opposed to the more specific "hogs," "sows," or "swine," except for when a specific example is called for. Moreover, "cattle" is a fraught term—in *Uncle Tom's Cabin*, enslaved people are sometimes referred to as "cattle," and it shares a root with "chattel." The word in this chapter refers to all things bovine—cows, mostly.
4. See Eric Sundquist, *To Wake the Nations*, 271–347.
5. Chesnutt, "Uncle Wellington's Wives," 223.
6. Indeed, pork/pig remains the most consumed meat in the world into the twenty-first century, more than beef/cattle (see USDA reports).
7. V. Anderson, *Creatures of Empire*, 110.
8. Stoll adds that Carolinians "never seemed to have recognized how slavery and cotton combined to frustrate restoration at almost every turn," which added up to "a brutal agricultural regime" and "a burdensome environment to till" (*Larding the Lean Earth*, 125).
9. Stoll, *Larding the Lean Earth*, 130–31.
10. For example, the state had 9,100,000 breeding pigs in 2018 and 75,179,000 chickens hatched in November 2018 alone. USDA, *Quarterly Hogs and Pigs*; USDA. *Chickens and Eggs*.
11. Westward expansion, of course, had begun while slavery was in practice in the United States. For more on the transportation of slaves from the eastern seaboard states west to states like Mississippi and Alabama, see Saunt, *Unworthy Republic*, 190–98.
12. Williams-Forson describes images like "postcards, sheet music, greeting cards, and other ephemera illustrating bandanna-wearing women and chicken-stealing men" (*Building Houses*, 16).
13. The ad itself "implies that African Americans can be distracted and pacified by a nicely cured piece of pig" from their very real socioeconomic struggles (Tompkins, *Racial Indigestion*, 177).
14. Eric Sundquist observes that "the best of [Chesnutt's] conjure and color line stories would frequently draw on the racist figures of popular culture in order to seize and invert their cultural power" (*To Wake the Nations*, 285).
15. Wallach, *Dethroning the Deceitful Pork Chop*, 170.
16. Wallach, *Dethroning the Deceitful Pork Chop*, 168.

17. Washington, *Up from Slavery*, 139.
18. Quoted in Wallach, *Dethroning the Deceitful Pork Chop*, 169.
19. For more on the issue of survivance, which was developed to describe Native American history and culture, see Powell, "Rhetorics of Survivance"; and Vizenor, *Survivance*.
20. Chesnutt, "Superstitions and Folk-lore of the South," 864.
21. Chesnutt, "Superstitions and Folk-lore of the South," 866.
22. J. Anderson, *Conjure in African American Society*, x.
23. Anderson adds: "For Chesnutt conjure was one means by which blacks could deal with the injustices of the ruling class. For white authors hoodoo symbolized black barbarism, a necessary counterpart to their conception of white civilization. For both races it was part of what it meant to be southern" (*Conjure in African American Society*, 9).
24. Buell, "Toxic Discourse," 657.
25. As we saw in chapter 5, Ruiz de Burton argues for a railroad to connect the South with southern California. That railroad plan is ultimately thwarted by the railroad monopolists, and we readers are left to speculate as to what the fates of the Californios and the small-scale farmers of the South would have been with such a connecting lifeline in the late nineteenth century.
26. Chesnutt is responding to a historical moment where mining and logging, in addition to land expansion and agricultural intensification, were increasing at unprecedented rates, which were to bring disastrous environmental consequences. According to Matthew Taylor, posthumanism might be seen as "disturbingly imperialistic," as it is "not our interiority that disappears but the externality of the world, which becomes merely the shadow of our now universalized selves" (*Universes without Us*, 7–8).
27. Omi and Winant, *Racial Formation*.
28. Outka, *Race and Nature*, 3, 104.
29. Outka writes: "Under slavery, whites frequently conflated African Americans with domesticated animals and pastoral agriculture." He adds: "But these initial, straightforward conjurations function as a sort of précis for a more disturbing conjuration, one that reveals more clearly the impossible traumatic scene of slavery that Chesnutt, via Julius, will return to obsessively" (*Race and Nature*, 7, 111).
30. I call the character "Julius," as McAdoo is the last name of his slave owner, and "Uncle" is an inappropriate epithet.

31. Gates and Tatar, *Annotated African American Folktales*, xlvi.
32. White, "From Globalized Pig Breeds," 94–97.
33. "Pigs and chickens were, of course, the easiest meat to carry to the New World by ship," note Root and de Rochemont, and thus, beginning in the 1600s, "pork was America's first important meat from domestic animals, and would remain the country's most common meat for three centuries" (*Eating in America*, 59).
34. Chesnutt, "Conjurer's Revenge," 25.
35. Root and de Rochemont, *Eating in America*, 68–69. There was also abundant corn production in the West in the nineteenth century, primarily for feed for cattle, pigs, and other livestock. Pigs came to be considered "fifteen or twenty bushels of corn on four legs" (White, "From Globalized Pig Breeds," 110). Adrienne Petty researches small-scale farms in the South, often cared for by emancipated Southerners, of the kind represented by the conjure man in this story. Into the early twentieth century, small-scale was considered "one cow, two pigs, 50 chickens, and a garden on each farm" (*Standing their Ground*, 91).
36. Chesnutt, "Conjurer's Revenge," 25.
37. Chesnutt, "Conjurer's Revenge," 26–27.
38. Wallach writes: "Beef consumption, it follows, could transform the eater from the inside, priming him or her for full incorporation into the US nation-state" (*Dethroning the Deceitful Pork Chop*, 168).
39. Chesnutt, "Conjurer's Revenge," 29.
40. With both the shoat and the beef, African Americans are seen carrying livestock and their meat on their backs throughout this story.
41. Chesnutt, "Conjurer's Revenge," 27.
42. Chesnutt, "Conjurer's Revenge," 27.
43. Hurston, *Mules and Men*, 22. In addition, Erik Redling observes the connection between Chesnutt's tales and *The Golden Ass* of Apuleius: "Bakhtin's assertion that the chronotope of *The Golden Ass* influenced the development of the European novel has to be seen as an understatement when applied to Chesnutt's dialect story 'The Conjurer's Revenge'" (*"Speaking of Dialect,"* 112).
44. Stepto and Greeson in Chesnutt, *Conjure Stories*, 24n.
45. Stepto, "The Cycle," 271. SallyAnn H. Ferguson calls the outcome of this story Julius's "only direct economic victory over John," adding that Chesnutt seems to want "to show, for once, that this same black is

'man' enough to part the white from his dollars and that he can do so alone" ("Chesnutt's 'The Conjurer's Revenge," 37–38).

46. Chesnutt, "Conjurer's Revenge," 31.

47. In his introduction to *The Conjure Stories*, Stepto notes that throughout the stories "Julius gets what he seems to want usually through Annie's intervention, and Annie's willingness to intervene is often predicated upon the extent to which she has become a full partner in the *event* of the telling of the tale" (xi).

48. Taylor observes that in the *Conjure Stories*, "if the frames are the realm of rational, conscious agents, then the tales mark the uncanny return of the repressed" (*Universes without Us*, 122). Stepto adds that for Primus "there is pitifully little difference between his working the fields as a mule and working them as a slave" ("The Cycle," 271).

49. There has been a great deal of controversy about Chesnutt's stance on the issue of "Why the Darkey Loves Chicken," the story's secondary title. Julius notes that all the African Americans in North Carolina are directly connected to that story, the characters in the story, the reason behind their love of chicken meat, and therefore to that plantation.

50. Chesnutt has a theme of Euro-American men suffering from their own greed through several of his *Conjure Stories*; "Goophered Grapevine" is another one.

51. Williams-Forson, *Building Houses*, 2.

52. Williams-Forson, *Building Houses*, 6. Also see her history of chicken (15–16).

53. In his history of beef in the United States, Specht observes: "Beef was an important social marker; it could not simply be replaced by chicken, fish, or, worst of all, a potato" (*Red Meat Republic*, 221).

54. Chesnutt, "A Victim of Heredity," 71.

55. For analysis of food, folktales, and racism in "Dave's Neckliss," see Swift and Mammoser, "Out of the Realm of Superstition"; and Fleissner, "Earth-Eating, Addiction, Nostalgia."

56. Chesnutt, "A Victim of Heredity," 71.

57. Washington, *Up from Slavery*, 5.

58. Chesnutt, "A Victim of Heredity," 76.

59. Chesnutt, "A Victim of Heredity," 75.

60. Root and de Rochemont reference "the famous 'hog and hominy' diet" prevalent in the South (*Eating in America*, 145). Chesnutt gives

other references to meat and bread in "Lonesome Ben," and the concept of rations appears again in "Mars Jeems's Nightmare," where Solomon gives the agricultural report quoted above.

61. Chesnutt, "A Victim of Heredity," 77–78.
62. Chesnutt, "A Victim of Heredity," 79.
63. Chesnutt, "A Victim of Heredity," 79.
64. Carpio, "Black Humor," 332.
65. Sundquist, *To Wake the Nations*, 381.
66. Glenda Carpio writes that in "A Victim of Heredity" "we see John briefly discover the human being behind the stereotype of the chicken thief." Ultimately, though, "rather than realize the connection between the exploitation of slaves and the exploitation of ex-slave employees like Julius, Annie is content to let people steal her 'portable property' and John would rather punish the 'victims of heredity' than examine his own complicity." ("Black Humor," 335, 332).
67. Carpio, "Black Humor," 332.
68. Sundquist, *To Wake the Nations*, 285.
69. Stepto notes, discussing the story's minstrel-like tone, that "the moments are not wholly comic" (Introduction, xvi).
70. Chesnutt also was a horse carriage driver. See Chesnutt, *"To Be an Author,"* 10.
71. The desire for beef, according to Wallach, "was certainly rooted in traditional English preferences. . . . [T]his form of animal protein soon became associated with civilization and whiteness" (*Dethroning the Deceitful Pork Chop*, 168). It is interesting, then, that beef was relatively rare in the diet of both Euro-American and African American southerners in the *Conjure Stories*.
72. Terry, *Afro-Vegan*, 1.
73. For more on this concept see Houston, "Making Do."

7. Industrial-Global Cattle

1. Sinclair, *The Jungle*, 38.
2. Eaton, *Tama*, 4.
3. Eaton, *Tama*, 15.
4. According to Donna Gabaccia, food reflects "cultural or social affinities in moments of change or transformation" (*We Are What We Eat*, 9). William Cronon adds that Chicago's "markets allowed people to look

farther and farther afield for the goods they consumed, vastly extending the distance between points of ecological production and points of economic consumption" (*Nature's Metropolis*, 266). For theories of the need for cosmopolitanism at times of ethnic violence and hostility in the United States, see Appiah, *Cosmopolitanism*; Gilroy, *Postcolonial Melancholia*; and Butler, *Precarious Life*.

5. As biographer Lauren Coodley notes of Sinclair's time researching the Union stockyards, "Nothing about this stockyard crammed with bawling cattle, nor this hotel populated exclusively by men, would have seemed familiar to him" (*Upton Sinclair*, 40). Sinclair remained near the stockyards for seven weeks, visiting Jane Addams's Hull House and interviewing the laborers of the area, realizing that "by carrying a dinner pail, he could go anywhere" (42).

6. Sinclair, "What Life Means to Me," 594.

7. Tavernier-Courbin, "*The Call of the Wild* and *The Jungle*," 250.

8. Jean Lee Cole observes that this period of Japonisme was not long-lived, however. After military success against China and Russia "reawakened fears of a Japanese yellow peril," anti-immigration laws were passed against the Japanese during the first two decades of the twentieth century (*The Literary Voices of Winnifred Eaton*, 48). In the years leading up to World War II, Eaton would go so far as to claim to "hate" her Japanese ethnicity: "I'm ashamed of having written about the Japanese, I hate them so" (qtd. in Ferens, *Edith and Winnifred Eaton*, 140).

9. Yuko Matsukawa notes that Watanna's pen name is also the name of a pen ("Cross-Dressing and Cross-Naming," 107–9).

10. Ling, "Winnifred Eaton," 8–10; Ferens, *Edith and Winifred Eaton*, 123.

11. Diana Birchall observes that Watanna's books were highly successful: *A Japanese Nightingale* (1902) was an "enormous best-seller," and her other books tended to sell between five thousand and forty thousand copies each (*Onoto Watanna*, 107). *A Japanese Nightingale* was reported to be one of the last books Olivia Clemens read during her life, according to the Mark Twain House library, as detailed on Donna Campbell's American literature website https://public.wsu.edu/~campbelld/amlit/watanna.htm.

12. Eaton, *Tama*, 169. Ironically, research has shown that Tama's diet is almost universally a healthier one than Tojin's.

13. Amy Ling notes that although Tama is blind, and an outcast, she has provided for herself for most of her life, and it is she "who shows a highly

respected American professor how to trap fish and where to find wild fruits and vegetables" ("Winnifred Eaton," 11).

14. Dominica Ferens notes that during this period, "Franz Boas had barely begun his lifelong project of cleaving the concept of 'culture' from that of 'race'" (*Edith and Winnifred Eaton*, 121). Gretchen Murphy adds that while Eaton shows the "tension in early formulations of Asian American cultural identity" in her work, she "nonetheless reinforced and drew from racisms underwritten by nationalist agendas for global power" (*Shadowing the White Man's Burden*, 184–85).

15. H. J. Dawson, "Winston Churchill and Upton Sinclair," 74.

16. H. J. Dawson, "Winston Churchill and Upton Sinclair," 75.

17. Sinclair, *The Jungle*, 47. Eric Schlosser writes that Sinclair's work "remains the essential starting point for an understanding of America's meatpacking industry today" (*Fast Food Nation*, 321n). He also summarizes Sinclair's most memorable scenes: "the routine slaughter of diseased animals, the use of chemicals such as borax and glycerine to disguise the smell of spoiled beef, the deliberate mislabeling of canned meat, the tendency of workers to urinate and defecate on the kill floor" (204). Cronon discusses the environmental impact of the Beef Trust: "Obsessed with turning waste into profit whatever the noneconomic cost, they sold what they should have thrown away—and yet did little to prevent pollution from the wastes that finally washed down their sewers" (*Nature's Metropolis*, 253).

18. Yoder, *Upton Sinclair*, 30.

19. For more on the Great Migration see Isabel Wilkerson's *The Warmth of Other Suns*.

20. Sinclair, *The Jungle*, 37.

21. Cronon observes that the Union stockyards were "capable of handling 21,000 head of cattle, 75,000 hogs, 22,000 sheep, and 200 horses, all at the same time" (*Nature's Metropolis*, 210). Ironically, Chicago "benefited more than any other city from the long-term effects of the animals' [bison's] annihilation" (218).

22. Sinclair, *The Jungle*, 42, 33.

23. Sinclair, *The Jungle*, 158–59.

24. Sinclair, *The Jungle*, 158.

25. Sinclair, *The Jungle*, 321.

26. Sinclair, *The Jungle*, 327.

27. Sinclair, *The Jungle*, 39.
28. Sinclair, *The Jungle*, 317.
29. Duvall, "Processes of Elimination," 29.
30. Sinclair, *The Jungle*, 408–9.
31. Sinclair, *The Jungle*, 409.
32. For more on the relationship between the Eaton sisters and their work, see Lim, "Sibling Hybridities."
33. Ling writes that Winnifred is not the only Eaton child to "pass" as non-Chinese. Siblings were "passing as Mexican, as English, as Japanese," but "Edith was the only one among fourteen children to choose resistance [being honest to her Chinese heritage], the more difficult and more noble path" ("Creating One's Self," 311).
34. Eaton, *Cattle*, 86.
35. Eaton, *Cattle*, 1.
36. Eaton, *Cattle*, 2.
37. Eaton, *Cattle*, 3.
38. Eaton, *Cattle*, 4.
39. Eaton, *Cattle*, 63. Eaton's dialect writing is somewhat difficult for modern readers.
40. Eaton, *Cattle*, 96.
41. Alexander Smith writes: "An estimated 20,000 to 30,000 Chinese from this area [Canton (Guangdong)] came to California during the gold rush" (*Eating History*, 47). David Y. H. Wu and Sidney C. H. Cheung observe: "For the sake of survival, Chinese immigrants in the frontier societies in the Western world provide local, peasant style, dishes in the restaurants to satisfy a global idea of authentic Chinese cuisine" (Introduction, 12). Indeed, Colleen Lye observes that "the mode of incorporation of Asian migrant labor into the United States forces us to grapple with the fully industrial modernity of race," starting "from the era when U.S. industrialization began to depend upon the importation of transnational Asian labor and capital expansion into the Asia-Pacific" (*America's Asia*, 9).
42. Following the completion of the transcontinental railroad in 1869, however, the need for an abundance of cheap labor in the United States came to a temporary halt. At that point the Chinese went from being actively recruited to a "yellow peril." At this point, adds Lisa Lowe, "Orientalist racializations of Asians as physically and intellectually different from 'whites' predominated especially in periods in which a domestic

crisis of capital was coupled with nativist anti-Asian backlash" (*Immigrant Acts*, 4–5). In 1882 Congress passed the Chinese Exclusion Act, the first of several such acts against Chinese immigration; others were passed and strengthened in 1882, 1884, 1886, and 1888. Erika Lee dates the "exclusion era" for the Chinese as from 1882 through 1943, when the act was finally repealed by the Magnuson Act (*At America's Gates*, 4). James G. Blaine, the secretary of state under President James Garfield, campaigned for these exclusion acts. In a letter to President Garfield on the subject, Blaine writes: "You will, I think, be compelled to take the ground that a servile class—assimilating in all its conditions of labor to chattel slavery—must be excluded from free immigration" (qtd. in Tyler, *Foreign Policy of James G. Blaine*, 18). Sherry Inness observes that such harsh tactics "caused the Chinese population in America to fall from 107,488 to 71,531" by 1910 (*Secret Ingredients*, 43). The Chinese during this time were subjected to some of the most virulent xenophobia in the nation's history.

43. Eaton, *Cattle*, 52. Lee's domesticity is contrasted to that of Mrs. Langdon, and later Nettie Day, which is more like the domesticity idealized by Susan Fenimore Cooper than the male-focused Chum Lee, as cook, and the cowboys. Lee did the "main cooking." But the women performed their own feats of domestic labor.

44. Eaton, *Cattle*, 278.

45. Eaton, *Cattle*, 279.

46. Lowe, *Immigrant Acts*, 12. Ronald Takaki adds that during this period Chinese could earn up to "thirty dollars a month" in California compared to "three to five dollars a month in China" (*A Different Mirror*, 179).

47. Lepore, *These Truths*, 334. Lepore also connects U.S. labor unions, of the kind described by Sinclair in *The Jungle*, as in competition with immigrants: "No group of native-born Americans was more determined to end Chinese immigration than factory workers" (*These Truths*, 336).

48. Eaton, *Cattle*, 291.

49. Eaton, *Cattle*, 293.

50. Shapiro, *Perfection Salad*, 201. She warns, however, that "scientific resistance was strong" to the introduction of exotic tastes to the national cuisine (201). Inness uses examples of Hispanic and Jewish communities to show that cuisine can be a significant stage in the acceptance of one people by another: "Hispanic cookbooks convey not only recipes, but

also spread pride in Hispanic culture and its accomplishments. Jewish cookbooks share recipes and keep traditions alive" (*Secret Ingredients*, 5). These cookbooks attempt to educate, "spread pride," and "keep traditions alive" through their recipes. "During a time known for extreme xenophobia," Inness adds, "Chinese cooking literature served as a bridge between cultures" (39).

51. Leonardi, "Recipes for Reading," 340.
52. The cookbook would be Eaton's most Chinese-focused work. She returns to the issue of diet once more in "Starving and Writing in New York."
53. Bosse and Watanna, *Chinese-Japanese Cook Book*, 2.
54. Bosse and Watanna, *Chinese-Japanese Cook Book*, 2.
55. Upton Sinclair, "My Life in Diet," 3–4, box 33, folder 46. Sinclair MSS, 1814–1965, Lilly Library, Indiana University, Bloomington.
56. Bray, *Rice*, 6.

Conclusion

1. Ozeki, *My Year of Meats*, 126.
2. USDA, USDA *Agricultural Projections to 2028*, 99. Sumner adds: "Big exporting regions include South America, Oceania, and North America. The big importing regions are East and Southeast Asia, North America, Europe, and Africa. The United States is a major importer of ground beef and a major exporter of muscle cuts of beef" ("Economics of Global Cattle Industries," 19).
3. USDA, USDA *Agricultural Projections to 2028*, 55.
4. Indeed, Australia was already developing extensive cattle production in the nineteenth century, in many ways comparable to the U.S. and Canadian Great Plains. Warren Elofson compares "the first cattle ranching communities in the Northern Territory of Australia" in terms of their comparable "ecological, geographic, and weather-related conditions" ("Universality of Frontier Disorder" 147–48).
5. Ozeki, *My Year of Meats*, 9–10.
6. Ozeki, *My Year of Meats*, 124–26.
7. Ozeki, *My Year of Meats*, 254.
8. For useful theorizing of food studies in literature, see Korsmeyer, *Making Sense of Taste*, especially 185–223; and Gigante, *Taste*.
9. Gabaccia, *We Are What We Eat*, 8.

10. For a useful companion history on nineteenth-century cattle production in the United States, see Specht, *Red Meat Republic*.

11. Ben Beaumont-Thomas, "Paul McCartney Calls for 'Medieval' Chinese Markets to Be Banned over Coronavirus," *The Guardian*, April 14, 2020, https://www.theguardian.com/music/2020/apr/14/paul-mccartney-calls -for-medieval-chinese-markets-to-be-banned-over-coronavirus; Steve Almasy and Raja Razek, "Colorado Meat Packing Plant with Thousands of Employees Closed after Coronavirus Outbreak," CNN, April 11, 2020, https://www.cnn.com/2020/04/10/us/colorado-meat-packing -plant-coronavirus/index.html; and Isis Almeida and Vincent del Giudice, "Hundreds of Meat Workers Have Now Tested Positive for Coronavirus," *Los Angeles Times*, April 10, 2020, https://www.latimes.com /business/story/2020-04-10/meat-workers-coronavirus.

12. Nestle, *Pet Food Politics*, 2.

BIBLIOGRAPHY

"Account of the Buffalo." *Massachusetts Magazine, or, Monthly Museum of Knowledge and Rational Entertainment.* April 1, 1792, p. 4. American Antiquarian Society.

Adamson, Joni. "Medicine Food: Critical Environmental Justice Studies, Native North American Literature, and the Movement for Food Sovereignty." *Environmental Justice* 4, no. 4 (December 2011): 213–19.

Ahmad, Diana L. *Success Depends on the Animals: Emigrants, Livestock, and Wild Animals on the Overland Trails, 1840–1869.* Reno: University of Nevada Press, 2016.

Alaimo, Stacy. *Undomesticated Ground: Recasting Nature as Feminist Space.* Ithaca NY: Cornell University Press, 2000.

Alemán, Jesse. "Citizenship Rights and Colonial Whites: The Cultural Work of María Amparo Ruiz de Burton's Novels." In *Complicating Constructions: Race, Ethnicity, and Hybridity in American Texts,* edited by David S. Goldstein and Audrey B. Thacker, 3–30. Seattle: University of Washington Press, 2011.

———. "Novelizing National Discourses: History, Romance, and Law in *The Squatter and the Don.*" In *Recovering the U.S. Hispanic Literary Heritage,* vol. 3, edited by María Herrera-Sobek and Virginia Sánchez Korrol, 38–49. Houston: Arte Público Press, 2000.

Allen, Paula Gunn. "The Sacred Hoop: A Contemporary Perspective." In *The Ecocriticism Reader,* edited by Cheryll Glofelty and Harold Fromm, 241–63. Athens: University of Georgia Press, 1995.

Anderson, Jeffrey. *Conjure in African American Society.* Baton Rouge: Louisiana State University Press, 2005.

Anderson, Virginia DeJohn. *Creatures of Empire: How Domestic Animals Transformed Early America.* New York: Oxford University Press, 2004.

Antelyes, Peter. *Tales of Adventurous Enterprise: Washington Irving and the Poetics of Western Expansion.* New York: Columbia University Press, 1990.

Appadurai, Arjun. "How to Make a National Cuisine: Cookbooks in Contemporary India." *Comparative Studies in Society and History* 30, no. 1 (January 1988): 3–24.

Appiah, Kwame Anthony. *Cosmopolitanism: Ethics in a World of Strangers.* Reprint. New York: Norton, 2007.

Aranda, José F, Jr. "Contradictory Impulses: María Amparo Ruiz de Burton, Resistance Theory, and the Politics of Chicano/a Studies." *American Literature* 7, no. 3 (September 1998): 551–79.

Bakhtin, Mikhail. *Rabelais and His World.* Translated by Helen Iswolsky. Bloomington: Indiana University Press, 1984.

Barthes, Roland. "Toward a Psychosociology of Contemporary Food Consumption." In *Food and Culture: A Reader,* edited by Carole Counihan and Penny Van Esterik, 28–35. 2nd ed. New York: Routledge, 2008.

Baym, Nina. *American Women of Letters and Nineteenth-Century Sciences: Styles of Affiliation.* New Brunswick NJ: Rutgers University Press, 2002.

Bidwell, Percy W. "The Agricultural Revolution in New England." *American Historical Review* 26, no. 4 (1921): 683–702.

Birchall, Diana. *Onoto Watanna: The Story of Winnifred Eaton.* Urbana: University of Illinois Press, 2001.

Bosse, Sara, and Onoto Watanna [Winnifred Eaton]. *Chinese-Japanese Cook Book.* New York: Rand McNally, 1914.

Botha, Marc. "Toward a Critical Poetics of Securitization: A Response to Anker, Castronovo, Harkins, Masterson, and Williams." *American Literary History* 28, no. 4 (Winter 2016): 779–86.

Bourdieu, Pierre. *Distinction: A Social Critique of the Judgment of Taste.* Translated by Richard Nice. Cambridge: Harvard University Press, 1984.

Bradley, Donna. *Native Americans of San Diego County.* San Francisco: Arcadia, 2009.

Bray, Francesca. *Rice: Global Networks and New Histories.* New York: Cambridge University Press, 2017.

Breitweiser, Michael. *National Melancholy: Mourning and Opportunity in Classic American Literature.* Stanford: Stanford University Press, 2007.

Bryant, William Cullen. "The Prairies." In *Poetical Works of William Cullen Bryant: Household Edition*, 130–33. New York: Appleton, 1880.

Buell, Lawrence. *The Environmental Imagination: Thoreau, Nature Writing, and the Formation of American Culture*. Cambridge: Harvard University Press, 1995.

——. "Thoreau and the Natural Environment." In *The Cambridge Companion to Henry David Thoreau*, edited by Joel Myerson, 171–93. New York: Cambridge University Press, 1995.

——. "Toxic Discourse." *Critical Inquiry* 24, no. 3 (Spring 1998): 639–65.

Burns, Mark K. "Ineffectual Chase: Indians, Prairies, Buffalo, and the Quest for the Authentic West in Washington Irving's *A Tour on the Prairies*." *Western American Literature* 42, no. 1 (Spring 2007): 55–79.

Butler, Judith. *Precarious Life: The Powers of Mourning and Violence*. London: Verso, 2004.

Cafaro, Phillip. "Thoreau on Science and System." *Paideia Archives*. 1998. http://www.bu.edu/wcp/Papers/Envi/EnviCafa.htm.

Calcaterra, Angela. *Literary Indians: Aesthetics and Encounter in American Literature to 1920*. Chapel Hill: University of North Carolina Press, 2018.

Campbell, Donna. "Onoto Watanna (Winifred Eaton)." *American Literature Site*. https://public.wsu.edu/~campbelld/amlit/watanna.htm.

Canfield, Gae Whitney. *Sarah Winnemucca of the Northern Paiutes*. Norman: University of Oklahoma Press, 1983.

Carpenter, Cari M. "Sarah Winnemucca Goes to Washington: Rhetoric and Resistance in the Capital City." *American Indian Quarterly* 40, no. 2 (Spring 2016): 87–108.

Carpenter, Cari M., and Carolyn Sorisio. Introduction to *The Newspaper Warrior: Sarah Winnemucca Hopkins's Campaign for American Indian Rights, 1864–1891*, edited by Cari M. Carpenter and Carolyn Sorisio, 1–30. Lincoln: University of Nebraska Press, 2015.

Carpio, Glenda. "Black Humor in the Conjure Stories." In *The Conjure Stories*, edited by Robert B. Stepto and Jennifer Rae Greeson, 329–38. New York: Norton, 2012.

Castillo, Ana. Introduction to María Amparo Ruiz de Burton, *The Squatter and the Don*, edited by Ana Castillo, xiii–xviii. New York: Modern Library, 2004.

Cella, Matthew J. C. "Disturbing Hunting Grounds: Negotiating Geocultural Change in the Western Narratives of George Gatlin and Washington

Irving." *Literature in the Early American Republic: Annual Studies on Cooper and His Contemporaries* 6 (2014): 63–83.

Chesnutt, Charles Waddell. "Conjurer's Revenge." Chesnutt, *Conjure Stories*, 23–32.

——. *The Conjure Stories*. Edited by Robert B. Stepto and Jennifer Rae Greeson. New York: Norton, 2012.

——. "Mars Jeems's Nightmare." Chesnutt, *Conjure Stories*, 90–102. New York: Norton, 2012.

——. "Superstitions and Folk-lore of the South." In *Chesnutt: Stories, Novels, and Essays*, edited by Werner Sollors, 864–71. New York: Library of America, 2002.

——. *"To Be an Author": Letters of Charles W. Chesnutt, 1889–1905.* Edited by Joseph R. McElrath Jr. and Robert C. Leitz III. Princeton NJ: Princeton University Press, 1997.

——. "Uncle Wellington's Wives." In *Chesnutt: Stories, Novels, and Essays*, edited by Werner Sollors, 206–38. New York: Library of America, 2002.

——. "A Victim of Heredity: Or, Why the Darkey Loves Chicken." Chesnutt, *Conjure Stories*, 71–80.

Christie, John Aldrich. *Thoreau as World Traveler*. New York: Columbia University Press, 1965.

Chvany, Peter A. "'Those Indians Are Great Thieves, I Suppose?': Historicizing the White Woman in *The Squatter and the Don*." In *White Women in Racialized Spaces: Imaginative Transformation and Ethical Action in Literature*, edited by Samina Najmi, Rajini Srikanth, and Elizabeth Ammons, 105–18. Albany: State University of New York Press, 2002.

Clasen, Kelly "Gender and Genre in Susan Fenimore Cooper's Nature Writing." In *Creative Insights: American Creative Nonfiction*, edited by Jay Ellis, 79–94. Ipswich MA: Gray House, 2015.

Cole, Jean Lee. *The Literary Voices of Winnifred Eaton: Redefining Ethnicity and Authenticity*. New Brunswick NJ: Rutgers University Press, 2002.

Contreras, Alicia. "'I'll Publish Your Cowardice All Over California': María Amparo Ruiz de Burton's *The Squatter and the Don* in the Age of Howells." *American Literary Realism* 49, no. 3 (Spring 2017): 210–25.

Coodley, Lauren. *Upton Sinclair: California Socialist, Celebrity Intellectual*. Lincoln: University of Nebraska Press, 2013.

Cooper, James Fenimore. *The Pioneers: Or the Sources of the Susquehanna; A Descriptive Tale*. Edited by James Franklin Beard. Albany: State University of New York Press, 1980.

―――. *The Prairie: A Tale*. Edited by James P. Elliott. Albany: State University of New York Press, 1985.

Cooper, Susan Fenimore. *Essays on Nature and Landscape*. Edited by Rochelle Johnson and Daniel Patterson. Athens: University of Georgia Press, 2002.

―――. *Rural Hours*. Edited by Rochelle Johnson and Daniel Patterson. Athens: University of Georgia Press, 1998.

Cronon, William. *Nature's Metropolis: Chicago and the Great West*. New York: Norton, 1991.

Dawson, Hugh J. "Winston Churchill and Upton Sinclair: An Early Review of *The Jungle*." *American Literary Realism 1870–1919* 24, no. 1 (Fall 1991): 72–78.

Dawson, Melanie V. "Ruiz de Burton's Emotional Landscape: Property and Feeling in *The Squatter and the Don*." *Nineteenth-Century Literature* 63, no. 1 (2008): 41–72.

de la Luz Montes, Amelia M. "*Es Necesario Mirar Bien*: Nineteenth-Century Letter Making and Novel Writing in the Life of María Amparo Ruiz de Burton." In *Recovering the U.S. Hispanic Literary Heritage*, vol. 3, edited by María Herrera-Sobek and Virginia Sánchez Korrol, 16–37. Houston: Arte Público Press, 2000.

―――. "María Amparo Ruiz de Burton Negotiates American Literary Politics and Culture." In *Challenging Boundaries: Gender and Periodization*, edited by Joyce W. Warren and Margaret Dickie, 202–25. Athens: University of Georgia Press, 2000.

Deloria, Philip J. *Playing Indian*. New Haven: Yale University Press, 1998.

DeWitt, Dave. *The Founding Foodies: How Washington, Jefferson, and Franklin Revolutionized American Cuisine*. Naperville IL: Sourcebooks, 2010.

Dickens, Charles. *American Notes for General Circulation*. Vol. 2. 3rd ed. London: Chapman & Hall, 1842.

Dippie, Brian W. *The Vanishing American: White Attitudes and U.S. Indian Policy*. Middletown CT: Wesleyan University Press, 1982.

Dolan, Kathryn C. *Beyond the Fruited Plain: Food and Agriculture in U.S. Literature, 1850–1905*. Lincoln: University of Nebraska Press, 2014.

―――. "Diet and Vegetarianism." In *Henry David Thoreau in Context*, edited by James S. Finley, 216–24. New York: Cambridge University Press, 2017.

Douglas, Mary. "Deciphering a Meal." *Daedalus* 101, no. 1 (Winter 1972): 61–81.

Duvall, J. Michael. "Processes of Elimination: Progressive-Era Hygienic Ideology, Waste, and Upton Sinclair's *The Jungle.*" *American Studies* 43, no. 3 (Fall 2002): 29–56.

Eaton, Winnifred. *Cattle.* New York: W. J. Watt, 1924.

———. *Tama.* New York: Harper & Brothers, 1910.

Elofson, Warren. "The Universality of Frontier Disorder: Northern Australia Viewed against the Northern Great Plains of North America." *Agricultural History* 88, no. 2 (Spring 2014): 147–74.

Eves, Rosalyn Collings. "Finding Place to Speak: Sarah Winnemucca's Rhetorical Practices in Disciplinary Spaces." *Legacy: A Journal of American Women Writers* 31, no. 1 (2014): 1–22.

Ferens, Dominika. *Edith and Winnifred Eaton: Chinatown Missions and Japanese Romances.* Urbana: University of Illinois Press, 2001.

Ferguson, SallyAnn H. "Chesnutt's 'The Conjurer's Revenge': The Economics of Direct Confrontation." *Obsidian* 7 (1981): 37–42.

Fernández-Armesto, Felipe. *Near a Thousand Tables: A History of Food.* New York: Free Press, 2003.

Finley, James. "*Who* Are We? *Where* Are We? Contact and Literary Navigation in *The Maine Woods.*" *ISLE: Interdisciplinary Studies in Literature and Environment* 19, no. 2 (Spring 2012): 336–55.

Fiskio, Janet. "Sauntering across the Border: Thoreau, Nabhan, and Food Politics." In *The Cambridge Companion to Literature and the Environment*, edited by Louise Westling, 136–51. New York: Cambridge University Press, 2014.

———. "Unsettling Ecocriticism: Rethinking Agrarianism, Place, and Citizenship." *American Literature* 84, no. 2 (June 2012): 301–25.

———. "Where Food Grows on Water: Food Sovereignty and Indigenous North American Literatures." In *The Routledge Companion to Native American Literature*, edited by Deborah L. Madsen, 238–48. New York: Routledge, 2016.

Fitzgerald, Amy J. *Animals as Food: (Re)connecting Production, Processing, Consumption, and Impacts.* East Lansing: Michigan State University Press, 2015.

Fleissner, Jennifer. "Earth-Eating, Addiction, Nostalgia: Charles Chesnutt's Diasporic Regionalism." *Studies in Romanticism* 49, no. 2 (Summer 2010): 313–36.

Franklin, Wayne. *James Fenimore Cooper: The Early Years*. New Haven: Yale University Press, 2007.

Gabaccia, Donna. *We Are What We Eat: Ethnic Food and the Making of Americans*. Cambridge: Harvard University Press, 1998.

Garrard, Greg. *Ecocriticism*. New York: Routledge, 2012.

Gates, Henry Louis, Jr., and Maria Tatar. *The Annotated African American Folktales*. New York: Liveright, 2017.

Germic, Stephen. "Land Claims, Natives, and Nativism: Susan Fenimore Cooper's Fealty to Place." *American Literature* 79, no. 3 (September 2007): 475–500.

Gianquitto, Tina. "The Noble Designs of Nature: God, Science, and the Picturesque in Susan Fenimore Cooper's *Rural Hours*." In *Susan Fenimore Cooper: New Essays on Rural Hours and Other Works*, edited by Rochelle Johnson and Daniel Patterson, 169–91. Athens: University of Georgia Press, 2001.

Gigante, Denise. *Taste: A Literary History*. New Haven: Yale University Press, 2005.

Gilbert, Deborah Anne. "Articulating Identity in Literary Struggles Over Land: Helen Hunt Jackson, Sarah Winnemucca Hopkins, and María Amparo Ruiz de Burton." PhD diss., Stony Brook University, 2004.

Gillman, Susan. "The Squatter, the Don, and the Grandissimes in Our America." In *Mixing Race, Mixing Culture: Inter-American Literary Dialogues*, edited by Monika Kaup and Debra J. Rosenthal, 140–59. Austin: University of Texas Press, 2002.

Gilroy, Paul. *Postcolonial Melancholia*. New York: Columbia University Press, 2006.

González, Marcial. *Chicano Novels and the Politics of Form: Race, Class, and Reification*. Ann Arbor: University of Michigan Press, 2009.

Gura, Philip F. "Thoreau's Maine Woods Indians: More Representative Men." *American Literature* 49, no. 3 (November 1977): 366–84.

———. "A Wild, Rank Place: Thoreau's *Cape Cod*." In *The Cambridge Companion to Henry David Thoreau*, 142–51. New York: Cambridge University Press, 1995.

Haas, Lisbeth. *Conquests and Historical Identities in California, 1769–1936*. Berkeley, University of California Press, 1995.

Hanrahan, Heidi M. "'[w]orthy the imitation of the whites': Sarah Winnemucca and Mary Peabody Mann's Collaboration." *MELUS: Multiethnic Literature of the United States* 38, no. 1 (Spring 2013): 119–36.

Hanssen, Jessica Allen. "Transnational Narrativity and Pastoralism in *The Sketch Book of Geoffrey Crayon, Gent.* by Washington Irving." *Transnational Literature* 9, no. 1 (November 2016): 1–12.

Haraway, Donna. *The Companion Species Manifesto*. Chicago: Prickly Paradigm Press, 2003.

———. *When Species Meet*. Minneapolis: University of Minnesota Press, 2008.

Harding, Walter. *The Days of Henry Thoreau: A Biography*. Princeton NJ: Princeton University Press, 1982.

Hardt, Michael, and Antonio Negri. *Empire*. Cambridge: Harvard University Press, 2001.

Heaton, John W. *The Shoshone-Bannocks: Culture and Commerce at Fort Hall, 1870–1940*. Lawrence: University of Kansas Press, 2005.

"Henry Leavitt Ellsworth." The Miriam and Ira D. Wallach Division of Art, Prints and Photographs: Print Collection, The New York Public Library Digital Collections. Accessed April 18, 2020. http://digitalcollections.nypl .org/items/510d47df-8f35-a3d9-e040-e00a18064a99.

Hittman, Michael. *Wovoka and the Ghost Dance*. Lincoln: University of Nebraska Press, 1997.

Hoffman, Daniel G. "Irving's Use of American Folklore in 'The Legend of Sleepy Hollow.'" *PMLA* 68, no. 3 (June 1953): 425–35.

Horsman, Reginald. *Race and Manifest Destiny: Origins of American Racial Anglo-Expansionism*. Reprint. Cambridge: Harvard University Press, 1981.

Houston, Lynn Marie. "'Making Do': Caribbean Foodways and the Economics of Postcolonial Literary Culture." *MELUS* 32, no. 4 (Winter 2004): 99–113.

Howe, LeAnn. "Tribalography: The Power of Native Stories." *Journal of Dramatic Theory and Criticism* 14, no. 1 (Fall 1999): 3–12.

Hurston, Zora Neale. *Mules and Men*. New York: HarperCollins, 1999.

Igler, David. *Industrial Cowboys: Miller and Lux and the Transformation of the Far West, 1850–1920*. Berkeley: University of California Press, 2001.

Inness, Sherry. *Secret Ingredients: Race, Gender, and Class at the Dinner Table*. New York: Palgrave Macmillan, 2006.

Irmschler, Christoph. "Susan Fenimore Cooper's Ecology of Reading." *Partial Answers: Journal of Literature and the History of Ideas* 12, no. 1 (January 2014): 41–61.

Irving, Washington. *The Adventures of Captain Bonneville*. Edited by Robert A. Rees and Alan Sandy. Boston: Twayne, 1977.

———. *Astoria; or, Anecdotes of an Enterprize beyond the Rocky Mountains.* Edited by Richard Dilworth Rust. Boston: Twayne, 1976.

———. *Crayon Miscellany.* Edited by Dahlia Kirby Terrell. Boston: Twayne, 1979.

———. *The Journals of Washington Irving.* Vol. 3. Edited by William P. Trent and George S. Hellman. Boston: Bibliophile Society, 1919.

———. *The Sketch Book of Geoffrey Crayon, Gent.* Edited by Haskell Springer. Boston: Twayne, 1978.

Isenberg, Andrew C. *The Destruction of the Bison: An Environmental History, 1750–1920.* New York: Cambridge University Press, 2000.

Iverson, Peter. *When Indians Became Cowboys: Native Peoples and Cattle Ranching in the American West.* Norman: University of Oklahoma Press, 1994.

Jackson, Helen Hunt. *Century of Dishonor.* Boston: Little, Brown, 1917.

Jacobs, Margaret D. "Mixed-Bloods, Mestizas, and Pintos: Race, Gender, and Claims to Whiteness in Helen Hunt Jackson's *Ramona* and María Amparo Ruiz de Burton's *Who Would Have Thought It?*" *Western American Literature* 36, no. 3 (Fall 2001): 212–32.

Jacoby, Karl. *Crimes against Nature: Squatters, Poachers, Thieves, and the Hidden History of American Conservation.* Berkeley, University of California Press, 2014.

Jaros, Peter. "Irving's *Astoria* and the Forms of Enterprise." *American Literary History* 30, no. 1 (Spring 2018): 1–28.

Jefferson, Thomas. *Notes on the State of Virginia.* Edited by William Peden. Chapel Hill: University of North Carolina Press, 1954.

Jentz, Paul. *Seven Myths of Native American History.* Indianapolis: Hackett, 2018.

Johnson, Kendall. "Caleb Cushing and the Corporate Romance of Free Trade Imperialism in Washington Irving's *Astoria.*" *Literature Compass* (2017): 1–11.

Johnson, Linck C. *Thoreau's Complex Weave: The Writing of "A Week on the Concord and Merimack Rivers," with the Text of the First Draft.* Charlottesville: University Press of Virginia, 1986.

Johnson, Rochelle L. *Passions for Nature: Nineteenth-Century America's Aesthetics of Alienation.* Athens: University of Georgia Press, 2009.

———. "*Walden, Rural Hours,* and the Dilemma of Representation." In *Thoreau's Sense of Place: Essays in American Environmental Writing,* edited by Richard J. Schneider, 179–93. Iowa City: University of Iowa Press, 2000.

Johnson, Rochelle, and Daniel Patterson. Introduction to *Susan Fenimore Cooper: Essays on Nature and Landscape*, edited by Rochelle Johnson and Daniel Patterson, xiii–xxxii. Athens: University of Georgia Press, 2002.

———. Introduction to Susan Fenimore Cooper, *Rural Hours*, edited by Rochelle Johnson and Daniel Patterson, ix–xxii. Athens: University of Georgia Press, 1998.

Jones, Brian Jay. *Washington Irving: An America Original*. New York: Arcade, 2008.

Jones, David. Introduction to Susan Fenimore Cooper, *Rural Hours*, edited by David Jones, xiii–xxxviii. Syracuse: Syracuse University Press, 1968.

Kaufman, Frederick. "Gut Reaction: The Enteric Terrors of Washington Irving." *Gastronomica* 3, no. 2 (2003): 41–49.

Keyser, Catherine. *Artificial Color: Modern Food and Racial Fictions*. New York: Oxford University Press, 2019.

Kilcup, Karen L. *Fallen Forests: Emotion, Embodiment, and Ethics in American Women's Environmental Writing, 1781–1924*. Athens: University of Georgia Press, 2013.

———. "Writing against Wilderness: María Amparo Ruiz de Burton's Elite Environmental Justice." *Western American Literature* 47, no. 4 (Winter 2013): 360–85.

Kime, Wayne R. "The Author as Professional: Washington Irving's 'Rambling Anecdotes' of the West." In *Critical Essays on Washington Irving*, edited by Ralph M. Alderman, 237–53. Boston: Hall, 1990.

Klingborg, Donald J. Introduction to *The Welfare of Cattle*, edited by Terry Engle, Donald J. Klingborg, and Bernard E. Rollin, 3–8. Boca Raton FL: Taylor & Francis, 2019.

Klotz, Sarah. "The Red Man Has Left No Mark Here: Graves and Land Claim in the Cooperian Tradition." *ESQ* 60, no. 3 (2014): 331–69.

Knowlton, Christopher. *Cattle Kingdom: The Hidden History of the Cowboy West*. New York: Houghton Mifflin, 2017.

Kohler, Michelle. "Sending Word: Sarah Winnemucca and the Violence of Writing." *Arizona Quarterly: A Journal of American Literature, Culture, and Theory* 69, no. 3 (Autumn 2013): 49–54.

Kolodny, Annette. *The Lay of the Land: Metaphor as Experience and History in American Life and Letters*. Chapel Hill: University of North Carolina Press, 1975.

Korsmeyer, Carolyn. *Making Sense of Taste: Food and Philosophy*. Cornell NY: Cornell University Press, 2002.

Krell, Alan. *The Devil's Rope: A Cultural History of Barbed Wire*. London: Reaktion Books, 2002.

Kucich, John J. "Lost in the Maine Woods: Thoreau, Joseph Nicolar, and the Penobscot World." *Concord Saunterer* 19–20 (2011): 22–52.

Kurlansky, Mark. *The Big Oyster: History on the Half Shell*. New York: Random House, 2007.

Lape, Noreen Grover. "'I Would Rather Be with My People, but Not to Live with Them as They Live': Cultural Liminality and Double Consciousness in Sarah Winnemucca Hopkins's *Life among the Piutes: Their Wrongs and Claims*." *American Indian Quarterly* 22, no. 3 (Summer 1998): 259–79.

Larkin, Edward. "Time, Empire, and Nation." In *Time and Literature*, edited by Thomas M. Allen, 210–24. New York: Cambridge University Press, 2018.

Lazo, Rodrigo. "The Trans-American Literature of Conquest and Exile, 1823–1885." In *The Cambridge Companion to Latina/o American Literature*, edited by John Morán González, 3–16. New York: Cambridge University Press, 2016.

Leach, Hadley. "Thoreau's Aphoristic Form." *Arizona Quarterly* 68, no. 3 (2012): 1–26.

Lee, Erika. *At America's Gates: Chinese Immigration during the Exclusion Era, 1882–1943*. Chapel Hill: University of North Carolina Press, 2003.

LeMenager, Stephanie. *Manifest and Other Destinies: Territorial Fictions of the Nineteenth-Century United States*. Lincoln: University of Nebraska Press, 2004.

———. "Trading Stories: Washington Irving and the Global West." *American Literary History* 15, no. 4 (Winter 2003): 683–708.

Leonardi, Susan J. "Recipes for Reading: Summer Pasta, Lobster à la Riseholme, and Key Lime Pie." *PMLA* 104, no. 3 (May 1989): 340–47.

Lepore, Jill. *These Truths: A History of the United States*. New York: Norton, 2018.

Lewis, Meriwether, and William Clark. *The Journals of Lewis and Clark*. Edited by Bernard DeVoto. New York: Houghton Mifflin, 1997.

Lim, Shirley Geok-lin. "Sibling Hybridities: The Case of Edith Eaton/Sui Sin Far and Winnifred Eaton/Onoto Watanna." *Life Writing*. 4, no. 1 (April 2007): 81–99.

Limerick, Patricia Nelson. *The Legacy of Conquest: The Unbroken Past of the American West*. New York: Norton, 1987.

Ling, Amy. "Creating One's Self: The Eaton Sisters." In *Reading the Literatures of Asian America*, edited by Shirley Geok-lin Lim, Amy Ling, and Elaine H. Kim, 305–18. Philadelphia: Temple University Press, 1992.

———. "Winnifred Eaton: Ethnic Chameleon and Popular Success." *MELUS* 11, no. 3 (Autumn 1984): 5–15.

Littlefield, Daniel. "Washington Irving and the American Indian." *American Indian Quarterly* 5, no. 2 (May 1979): 135–54.

López, Marissa K. "Feeling Mexican: Ruiz de Burton's Sentimental Railroad Fiction." In *The Latino Nineteenth Century*, edited by Rodrigo Lazo and Jesse Alemán, 168–90. New York: New York University Press, 2016.

Lowe, Lisa. *Immigrant Acts: On Asian American Cultural Politics*. Durham NC: Duke University Press, 1996.

Luis-Brown, David. "'White Slaves' and the 'Arrogant Mestiza': Reconfiguring Whiteness in *The Squatter and the Don* and *Ramona*." *American Literature* 69, no. 4 (December 1997): 813–39.

Lye, Colleen. *America's Asia: Racial Form and American Literature, 1893–1945*. Princeton NJ: Princeton University Press, 2005.

Maddox, Lucy. "Susan Fenimore Cooper's Rustic Primer." In *Susan Fenimore Cooper: New Essays on Rural Hours and Other Works*, edited by Rochelle Johnson and Daniel Patterson, 83–95. Athens: University of Georgia Press, 2001.

Magee, Richard M. "Sentimental Ecology: Susan Fenimore Cooper's *Rural Hours*." In *Such News of the Land: U.S. Women Nature Writers*, edited by Thomas S. Edwards and Elizabeth A. DeWolfe, 27–36. Hanover NH: University Press of New England, 2001.

Mann, Charles C. *1491: New Revelations of the Americas before Columbus*. New York: Knopf, 2005.

Marx, Leo. *The Machine in the Garden*. New York: Oxford University Press, 1964.

Matsukawa, Yuko. "Cross-Dressing and Cross-Naming: Decoding Onoto Watanna." In *Tricksterism in Turn-of-the-Century American Literature: A Multicultural Perspective*, edited by Elizabeth Ammons and Annette White-Parks, 106–35. Hanover NH: University Press of New England, 1994.

May, Chad T. "The Romance of America: Trauma, National Identity, and the Leather-Stocking Tales." *Early American Studies: An Interdisciplinary Journal* 9, no. 1 (Winter 2011): 167–86.

McGann, Jerome. "Washington Irving, *A History of New York*, and American History." *Early American Literature* 47, no. 2 (2012): 349–76.

McKibben, Bill. "Nature and Environment." *The Atlantic*, May 2006. https://www.theatlantic.com/magazine/archive/2006/05/nature-environment/304780/.

McPhee, John. *The Founding Fish*. New York: Farrar, Strauss and Giroux, 2002.

McWilliams, Carey. *North from Mexico: The Spanish-Speaking People of the United States*. 3rd ed. Santa Barbara CA: ABC-CLIO, 2016.

Mead, Margaret. "The Problem of Changing Food Habits." In *Food and Culture: A Reader*, edited by Carole Counihan and Penny Van Esterik, 17–27. 2nd ed. New York: Routledge, 2008.

Mielke, Laura L. *Moving Encounters: Sympathy and the Indian Question in Antebellum Literature*. Amherst: University of Massachusetts Press, 2008.

Moldenhauer, Joseph J. "Historical Introduction." In Henry David Thoreau, *Cape Cod*, 249–96. Princeton NJ: Princeton University Press, 1988.

Moore, David L. *That Dream Shall Have a Name: Native Americans Rewriting America*. Lincoln: University of Nebraska Press, 2013.

Morgan, John. *New World Irish: Notes on One Hundred Years of Lives and Letters in American Culture*. New York: Palgrave Macmillan, 2011.

Morrison, Ronald. "Time, Place, and Culture in Thoreau's *Cape Cod*." *ESQ* 42, no. 3 (1996): 215–31.

Murphy, Gretchen. *Shadowing the White Man's Burden: U.S. Imperialism and the Problem of the Color Line*. New York: New York University Press, 2010.

Murray, Laura J. "The Aesthetic of Dispossession: Washington Irving and Ideologies of (De)Colonization in the Early Republic." *American Literary History* 8, no. 2 (Summer 1996): 205–31.

Myers, Jeffrey. *Converging Stories: Race, Ecology, and Environmental Justice in American Literature*. Athens: University of Georgia Press, 2005.

Nabhan, Gary Paul. *Cultures of Habitat: On Nature, Culture, and Story*. Washington DC: Counterpoint, 1997.

Neely, Michelle. "Embodied Politics: Antebellum Vegetarianism and the Dietary Economy of Walden." *American Literature* 85, no. 1 (March 2013): 33–60.

Nelson, Barbara "Barney." "Rustling Thoreau's Cattle: Wildness and Domesticity in 'Walking.'" In *Thoreau's Sense of Place: Essays in American Environmental Writing*, edited by Richard J. Schneider, 254–65. Iowa City: University of Iowa Press, 2000.

Nestle, Marion. *Pet Food Politics: The Chihuahua in the Coal Mine.* Berkeley: University of California Press, 2008.

Niemeyer, Mark. "From Savage to Sublime (And Partway Back): Indians and Antiquity in Early Nineteenth-Century American Literature." *Transatlantica* 2 (2015): 1–25.

Omi, Michael, and Howard Winant. *Racial Formation in the United States.* 3rd ed. New York: Routledge, 2015.

Outka, Paul. *Race and Nature from Transcendentalism to the Harlem Renaissance.* 2nd ed. New York: Palgrave Macmillan, 1994.

Ozeki, Ruth L. *My Year of Meats.* New York: Penguin Books, 1999.

Pellow, David Naguib, and Robert J. Brulle. "Power, Justice, and the Environment: Toward Critical Environmental Justice Studies." In *Power, Justice, and the Environment: A Critical Appraisal of the Environmental Justice Movement,* edited by David Naguib Pellow and Robert J. Brulle, 1–22. Cambridge: MIT Press, 2005.

Percy, Corinna Barrett. "Reformers, Racism, and Patriarchy in María Amparo Ruiz de Burton's *The Squatter and the Don.*" *Explicator* 75, no. 2 (2017): 112–17.

Pérez, Vincent. *Remembering the Hacienda: History and Memory in the Mexican American Southwest.* College Station: Texas A&M University Press, 2006.

Petersheim, Steven. "History and Place in the Nineteenth Century: Irving and Hawthorne in Westminster Abbey." *College Literature: A Journal of Critical Literary Studies* 39, no. 4 (2012): 118–37.

Petty, Adrienne Montieth. *Standing Their Ground: Small Farmers in North Carolina since the Civil War.* New York: Oxford University Press, 2013.

Pinsky, Robert. "Comedy, Cruelty, and Tourism: Thoreau's *Cape Cod.*" *American Scholar* 73, no. 3 (Summer 2004): 79–88.

———. Introduction to Henry David Thoreau, *Cape Cod,* edited by Joseph J. Moldenhauer, ix–xxiii. Princeton NJ: Princeton University Press, 2004.

Powell, Malea. "Princess Sarah, the Civilized Indian: The Rhetoric of Cultural Literacies in Sarah Winnemucca Hopkins's *Life among the Piutes.*" In *Rhetorical Women: Roles and Representations,* edited by Hildy Miller and Lillian Bidwell-Bowles, 63–80. Tuscaloosa: University of Alabama Press, 2005.

———. "Rhetorics of Survivance: How American Indians Use Writing." *College Composition and Communication* 53, no. 3 (February 2002): 396–434.

Ramirez, Pablo A. "American Imperialism in the Age of Contract: Herbert Spencer and the Defeat of Contractual Capitalism in Ruiz de Burton's *The Squatter and the Don.*" *ESQ* 57, no. 4 (2011): 427–55.

Redling, Erik. *"Speaking of Dialect": Translating Charles W. Chesnutt's "Conjure Tales" into Postmodern Systems of Signification*. Text & Theory Bd. 5. Königshausen & Neumann, 2006.

Reed, T. V. *The Art of Protest: Culture and Activism from the Civil Rights Movement to the Streets of Seattle*. Minneapolis: University of Minnesota Press, 2005.

Richardson, Robert D., Jr. *Henry Thoreau: A Life of the Mind*. Berkeley: University of California Press, 1986.

Rifkin, Mark. "Finding Voice in Changing Times: The Politics of Native Self-Representation during the Periods of Removal and Allotment." In *The Routledge Companion to Native American Literature*, edited by Deborah L. Madsen, 146–56. New York: Routledge, 2016.

Rimas, Andrew, and Evan D. G. Fraser. *Beef: The Untold Story of How Milk, Meat, and Muscle Shaped the World*. New York: HarperCollins, 2008.

Rivera, John-Michael. *The Emergence of Mexican America: Recovering Stories of Mexican Peoplehood in U.S. Culture*. New York: New York University Press, 2006.

Robinson, Kenneth Allen. *Thoreau and the Wild Appetite*. Hanover NH: Westholm, 1957.

Rodier, Katharine. "Authorizing Sarah Winnemucca? Elizabeth Peabody and Mary Peabody Mann." In *Reinventing the Peabody Sisters*, edited by Monika M. Elbert, Julie E. Hall, and Katharine Rodier, 108–25. Iowa City: University of Iowa Press, 2006.

Rogin, Michael Paul. *Fathers and Children: Andrew Jackson and the Subjugation of the American Indian*. New York: Knopf, 1975.

Root, Waverley, and Richard de Rochemont. *Eating in America: A History*. New York: HarperCollins, 1981.

Rose, Susan Dvorak. "Tracking the Moccasin Print: A Descriptive Index to Henry David Thoreau's Indian Notebooks and a Study of the Relationship of the Indian Notebooks to Mythmaking in *Walden*." PhD diss., University of Oklahoma, 1994.

Rubin-Dorsky, Jeffrey. *Adrift in the Old World: The Psychological Pilgrimage of Washington Irving*. Chicago: University of Chicago Press, 1998.

Ruiz de Burton, María Amparo. *The Squatter and the Don.* Edited by Ana Castillo. New York: Modern Library, 2004.

Ruoff, A. Lavonne Brown. "Early Native American Women Authors: Jane Johnston Schoolcraft, Sarah Winnemucca, Alice Callahan, E. Pauline Johnson, and Zitkala-Ša." In *Nineteenth-Century American Women Writers: A Critical Reader,* edited by Karen L. Kilcup, 81–111. Malden MA: Blackwell, 1998.

Saldívar, José David. "Nuestra América's Borders: Remapping American Cultural Studies." In *José Martí's "Our America": From National to Hemispheric Cultural Studies,* edited by Jeffrey Belnap and Raúl Fernández, 145–76. Durham NC: Duke University Press, 1998.

Sánchez, Rosaura, and Beatrice Pita, eds. *Conflicts of Interest: The Letters of María Amparo Ruiz de Burton.* Houston: Arte Público Press, 2001.

———. Introduction to María Amparo Ruiz de Burton, *The Squatter and the Don,* edited by Rosaura Sánchez and Beatrice Pita, 7–47. 2nd ed. Houston: Arte Público Press, 1997.

Saunt, Claudio. *Unworthy Republic: The Dispossession of Native Americans and the Road to Indian Territory.* New York: Norton, 2020.

Sayre, Robert F. *Thoreau and the American Indians.* Princeton NJ: Princeton University Press, 1977.

Schachterle, Lance. "On *The Prairie.*" In *Leather-Stocking Redux; or, Old Tales, New Essays,* edited by Jeffrey Walker, 124–49. New York: AMS Press, 2009.

Scherer, Joanna Cohan. "The Public Faces of Sarah Winnemucca." *Cultural Anthropology* 3, no. 2 (May 1998): 178–204.

Schlosser, Eric. *Fast Food Nation: The Dark Side of the All-American Meal.* New York: Mariner Books, 2012.

Schlueter, John P. "Private Practices: Washington Irving, Ralph Waldo Emerson, and the Recovery of Possibility." *Nineteenth-Century Literature* 66, no. 3 (December 2011): 283–306.

Schneider, Ryan. "Drowning the Irish: Natural Borders and Class Boundaries in Henry David Thoreau's *Cape Cod.*" *ATQ* 22, no. 3 (1987): 463–76.

Schnell, Michael. "'The Tasteful Traveller': Irving's *The Sketch Book* and Gustatory Self." *Mosaic* 35, no. 2 (2002): 111–27.

Senier, Siobhan. *Voices of American Indian Assimilation and Resistance: Helen Hunt Jackson, Sarah Winnemucca, and Victoria Howard.* Norman: University of Oklahoma Press, 2001.

Shapiro, Laura. *Perfection Salad: Women and Cooking at the Turn of the Century.* Berkeley: University of California Press, 2008.

Shields, Juliet. "Savage and Scott-ish Masculinity in *The Last of the Mohicans* and *The Prairie*: James Fenimore Cooper and the Diasporic Origins of American Identity." *Nineteenth-Century Literature* 64, no. 2 (September 2009): 137–62.

Shour, Nancy C. "Heirs to the Wild and Distant Past: Landscape and Historiography in James Fenimore Cooper's *The Pioneers.*" *James Fenimore Cooper Society Miscellaneous Papers* 10 (August 1998): 17–23.

Simon, Katie. "Affect and Cruelty in the Atlantic System: The Hauntological Argument of Henry David Thoreau's *Cape Cod.*" *ESQ* 62, no. 2 (2016): 245–82.

Sinclair, Upton. *The Jungle*. New York: Doubleday, 1906.

———. "What Life Means to Me." *Cosmopolitan Magazine*, October 1906, 591–95.

Sivils, Matthew Wynn. "Doctor Bat's Ass: Buffon, American Degeneracy, and Cooper's *The Prairie.*" *Western American Literature* 44, no. 4 (Winter 2010): 342–61.

Sizemore, Michelle R. "'Changing by Enchantment': Temporal Convergence, Early National Comparisons, and Washington Irving's Sketchbook." *Studies in American Fiction* 40 no. 2 (Fall 2013): 157–83.

Slatta, Richard W. *Cowboys of the Americas*. New Haven: Yale University Press, 1990.

Slotkin, Richard. *Regeneration through Violence: The Mythology of the American Frontier, 1600–1860*. Norman: University of Oklahoma Press, 2000.

Smalley, Andrea. *Wild by Nature: North American Animals Confront Colonization*. Baltimore: Johns Hopkins University Press, 2017.

Smith, Alexander. *Eating History: 30 Turning Points in the Making of American Cuisine*. New York: Columbia University Press, 2009.

Smith, Henry Nash. Introduction to James Fenimore Cooper, *The Prairie: A Tale*, edited by Henry Nash Smith, v–xxii. San Francisco: Rinehart Press, 1950.

———. *Virgin Land: The American West as Symbol and Myth*. Cambridge: Harvard University Press, 1978.

Sneider, Leah. "Gender, Literacy, and Sovereignty in Winnemucca's *Life among the Piutes.*" *American Indian Quarterly* 36, no. 3 (Summer 2012): 257–87.

Sorisio, Carolyn. "Playing the Indian Princess? Sarah Winnemucca's Newspaper Career and Performance of American Indian Identities." *Studies in American Indian Literatures* 23, no. 1 (Spring 2011): 1–37.

Specht, Joshua. *Red Meat Republic: A Hoof-to-Table History of How Beef Changed America*. Princeton NJ: Princeton University Press, 2019.

Specq, François, and Laura Dassow Walls. Introduction to *Thoreauvian Modernities: Transatlantic Conversations on an American Icon*, edited by François Specq, Laura Dassow Walls, and Michel Granger, 1–17. Athens: University of Georgia Press, 2013.

Spence, Mark David. *Dispossessing the Wilderness: Indian Removal and the Making of the National Parks*. New York: Oxford University Press, 1999.

Stark, Peter. *Astoria: John Jacob Astor and Thomas Jefferson's Lost Pacific Empire*. New York: HarperCollins, 2014.

Stepto, Robert B. "The Cycle of the First Four Stories." In *The Conjure Stories*, edited by Robert B. Stepto and Jennifer Rae Greeson, 266–83. New York: Norton, 2012.

———. Introduction to Charles Chesnutt, *The Conjure Stories*, edited by Robert B. Stepto and Jennifer Rae Greeson, vii–xxvii. New York: Norton, 2012.

Stoll, Steven. *Larding the Lean Earth: Soil and Society in Nineteenth-Century America*. New York: Hill and Wang, 2002.

Stromberg, Ernest. "Rhetoric and American Indians: An Introduction." In *American Indian Rhetorics of Survivance: Word Medicine, Word Magic*, edited by Ernest Stromberg, 1–12. Pittsburgh: University of Pittsburgh Press, 2006.

Sumner, Daniel A. "Economics of Global Cattle Industries, with Implications for On-Farm Practices." In *The Welfare of Cattle*, edited by Terry Engle, Donald J. Klingborg, and Bernard E. Rollin, 15–29. Boca Raton FL: Taylor & Francis, 2019.

Sundquist, Eric. *To Wake the Nations: Race in the Making of American Literature*. Cambridge MA: Belknap Press, 1993.

Sutton, John L. "'Savory Bison Hump': Carving Up American Identities in James Fenimore Cooper's *The Prairie*." In *The Image of the American West in Literature, the Media, and Society*, edited by Will Wright and Steven Kaplan, 271–74. Pueblo CO: Society for the Interdisciplinary Study of Social Imagery, 1996.

Sweet, Timothy. "Global Cooperstown: Taxonomy, Biogeography, and Sense of Place in Susan Fenimore Cooper's *Rural Hours*." *ISLE* 17, no. 3 (Summer 2010): 541–67.

Swift, John N., and Gigen Mammoser. "'Out of the Realm of Superstition': Chesnutt's 'Dave's Neckliss' and the Curse of Ham." *American Literary Realism* 42, no. 1 (Fall 2009): 1–12.

Szeghi, Tereza M. "The Vanishing Mexicana/o: (Dis)Locating the Native in Ruiz de Burton's *Who Would Have Thought It?* and *The Squatter and the Don.*" *Aztlán: A Journal of Chicano Studies* 36, no. 2 (Fall 2011): 89–120.

Takaki, Ronald. *A Different Mirror: A History of Multicultural America.* Rev. ed. New York: Back Bay Books, 2008.

Talley, Sharon. "Thoreau's Taste for the Wild in *Cape Cod.*" *Nineteenth-Century Prose* 31, no. 1 (Spring 2004): 82–100.

Tavernier-Courbin, Jacqueline. "*The Call of the Wild* and *The Jungle*: Jack London's and Upton Sinclair's Animal and Human Jungles." In *The Cambridge Companion to American Realism and Naturalism: Howells to London,* edited by Donald Pizer, 236–62. Cambridge: Cambridge University Press, 1995.

Taylor, Matthew. *Universes without Us: Posthuman Cosmologies in American Literature.* Minneapolis: University of Minnesota Press, 2013.

Terry, Bryant. *Afro-Vegan: Farm-Fresh African, Caribbean, and Southern Flavors Remixed.* New York: Ten Speed Press, 2014.

Thomas, Brook. *The Literature of Reconstruction: Not in Plain Black and White.* Baltimore: Johns Hopkins University Press, 2017.

———. "Ruiz de Burton, Railroads, Reconstruction." ELH 80, no. 3 (Fall 2013): 871–95.

Thoreau, Henry David. *Cape Cod.* Princeton NJ: Princeton University Press, 2004.

———. *Journal, Vol. 7, 1853–1854.* Edited by Nancy Craig Simmons and Ron Thomas. Princeton NJ: Princeton University Press, 2009.

———. *Journal XI: July 2, 1858–February 28, 1859.* Edited by Bradford Torrey. Boston: Houghton Mifflin, 1906.

———. *The Maine Woods.* Princeton NJ: Princeton University Press, 2004.

———. *Walden.* Princeton NJ: Princeton University Press, 2004.

———. *A Week on the Concord and Merrimack Rivers.* Princeton NJ: Princeton University Press, 2004.

Tompkins, Kyla Wazana. *Racial Indigestion: Eating Bodies in the Nineteenth Century.* New York: New York University Press, 2012.

Traisnel, Antoine. "American Entrapments: Taxonomic Capture in James Fenimore Cooper's *The Prairie.*" *Novel: A Forum on Fiction* 49, no. 1 (May 2016): 26–48.

Turner, Frederick Jackson. *History, Frontier, and Section: Three Essays.* Introduction by Martin Ridge. Albuquerque: University of New Mexico Press, 1993.

————. *The United States, 1830–1850: The Nation and Its Sections.* Berkeley: University of California Press, 1935.

Tuttle, Jennifer. "The Symptoms of Conquest: Race, Class, and the Nervous Body in *The Squatter and the Don.*" In *María Amparo Ruiz de Burton: Critical and Pedagogical Perspectives,* edited by Amelia María de la Luz Montes, 56–72. Lincoln: University of Nebraska Press, 2004.

Tyler, Alice Felt. *The Foreign Policy of James G. Blaine.* Hamden CT: Archon Books, 1965.

United Nations. "United Nations Declaration on the Rights of Indigenous Peoples." September 13, 2007. https://www.un.org/development/desa/indigenouspeoples /wp-content/uploads/sites/19/2018/11/UNDRIP_E_web.pdf.

USDA. *Chickens and Eggs.* December 21, 2018. https://downloads.usda.library .cornell.edu/usda-esmis/files/fb494842n/xs55mh49c/2227mt879/ckeg1218.pdf.

————. *Quarterly Hogs and Pigs.* December 20, 2018. https://downloads .usda.library.cornell.edu/usda-esmis/files/rj430453j/bc386p647/rf55zc904 /hgpg1218.pdf.

————. USDA *Agricultural Projections to 2028.* June 20, 2019. https://www .ers.usda.gov/webdocs/outlooks/92600/oce-2019-1.pdf?v=974.

Van Ballenberghe, Victor. *In the Company of Moose.* 2nd ed. Mechanicsburg PA: Stackpole, 2013.

Vance, William "'Man and Beast': The Meaning of Cooper's *The Prairie.*" *PMLA* 89, no. 2 (March 1974): 323–31.

Vizenor, Gerald. "Aesthetics of Survivance: Literary Theory and Practice." In *Survivance: Narratives of Native Presence,* edited by Gerald Vizenor, 1–23. Lincoln: University of Nebraska Press, 2008.

————. *Manifest Manners: Narratives on Postindian Survivance.* Lincoln: University of Nebraska Press, 1999.

————. *Survivance: Narratives of Native Presence.* Lincoln: University of Nebraska Press, 2008.

Wagner, Frederick. "Dining Out with Henry Thoreau: A Field Guide to Food Wild, Cultivated, and Transcendental." *Thoreau Society Bulletin* 175 (Spring 1986): 1–4.

Walker, Cheryl. *Indian Nation: Native American Literature and Nineteenth-Century Nationalisms.* Durham NC: Duke University Press, 1997.

Wallach, Jennifer Jensen. *Dethroning the Deceitful Pork Chop: Rethinking African American Foodways from Slavery to Obama.* Fayetteville: University of Arkansas Press, 2015.

Walls, Laura Dassow. *Henry David Thoreau: A Life*. Chicago: University of Chicago Press, 2017.

Warford, Elisa. "'An Eloquent and Impassioned Plea': The Rhetoric of Ruiz de Burton's *The Squatter and the Don*." *Western American Literature* 44, no. 1 (Spring 2009): 5–21.

Warren, Louis S. *God's Red Son: The Ghost Dance Religion and the Making of Modern America*. New York: Basic Books, 2017.

Washington, Booker T. *Up from Slavery: An Autobiography*. New York: Doubleday, 1901.

Watanna, Onoto. "Starving and Writing in New York." *Maclean's Magazine*, October 15, 1922, 66–67.

Watson, David. "Lawless Intervals: Washington Irving's *Astoria* and the Procession of Empire." *American Studies in Scandinavia* 42, no. 1 (2010): 5–24.

Weinstein, Josh A. "Susan Cooper's Humble Ecology: Humility and Christian Stewardship in *Rural Hours*." *Journal of American Culture* 35, no. 1 (March 2012): 65–77.

White, Sam. "From Globalized Pig Breeds to Capitalist Pigs: A Study in Animal Cultures and Evolutionary History." *Environmental History* 16 (January 2011): 94–120.

Wilkerson, Isabel. *The Warmth of Other Suns: The Epic Story of America's Great Migration*. New York: Random House, 2010.

Williams-Forson, Psyche A. *Building Houses out of Chicken Legs: Black Women, Food, and Power*. Chapel Hill: University of North Carolina Press, 2006.

Williamson, William D. *The History of the State of Maine: From Its First Discovery, A.D. 1602, to the Separation, A.D. 1820, Inclusive*. Hallowell ME: Glazer, Masters, 1832.

Wingate, Jordan. "Irving's Columbus and Hemispheric American History." *American Literature* 89, no. 3 (September 2017): 463–96.

Winnemucca Hopkins, Sarah. *Life among the Piutes: Their Wrongs and Claims*. Edited by Mary Peabody Mann. New York: Putnam's, 1883.

———. *The Newspaper Warrior: Sarah Winnemucca Hopkins's Campaign for American Indian Rights, 1864–1891*. Edited by Cari M. Carpenter and Carolyn Sorisio. Lincoln: University of Nebraska Press, 2015.

Wolfe, Cary. *Before the Law: Humans and Other Animals in a Biopolitical Frame*. Chicago: University of Chicago Press, 2013.

Wu, David Y. H., and Sidney C. H. Cheung. "Introduction: The Globalization of Chinese Food and Cuisine: Markers and Breakers of Cultural Barriers."

In *The Globalization of Chinese Food*, edited by David Y. H. Wu and Sidney C. H. Cheung, 1–18. Honolulu: University of Hawai'i Press, 2002.

Ybarra, Priscilla. *Writing the Goodlife: Mexican American Literature and the Environment.* Tucson: University of Arizona Press, 2016.

Yoder, Jon A. *Upton Sinclair.* New York: Ungar, 1975.

Zanjani, Sally. *Sarah Winnemucca.* Lincoln: University of Nebraska Press, 2001.

INDEX

Page numbers in italics refer to illustrations.

cattle (*cont.*)
    production, 2, 19, 48, 80,
    178, 198, 204, 209, 221, 236,
    279n4; ranches, 4, 32, 114–15,
    120–23, 126, 128–29, 132, 135,
    140, 143, 153, 158, 205, 209,
    218, 232; ranchers, 19, 112, 121,
    124–28, 131–32, 219–20, 223,
    269n68; ranching, 16, 123,
    139, 145, 165, 216, 258–59n7,
    261n23, 279n4; rustlers, 164,
    219; Spanish, 114; stockyards,
    19, 205, 206, 213–14, 220, 223,
    275n5, 276n21; as symbol, 4, 13,
    17, 230, 232–33, 235; as symbol
    in Chesnutt, 199, 202; as sym-
    bol in Eaton, 223; as symbol
    in Irving, 24, 34, 42–43, 49;
    as symbol in Ruiz de Burton,
    140, 144, 172; as symbol in
    Winnemucca, 110, 112–14, 120–
    23, 131, 133–34, 138, 260n16;
    territory, 122
Cattle (Eaton), 19, 205, 206, 215,
    216–24, 217
Century of Dishonor (Jackson), 121.
    See also Jackson, Helen Hunt
Channing, William Ellery, Jr., 80,
    102
Chesnutt, Andrew Jackson (father
    of Charles), 177
Chesnutt, Charles, 3, 17–18, 163,
    201, 202, 211, 233; agriculture
    and cuisine in, 36, 172–73,
    175–77, 193, 198–200; and beef,
    176, 186–88, 195–96, 199; and
    chicken, 179, 191–92, 195; and

conjure, 181–82, 190–91, 194,
    198, 271n23; and dialect writing,
    173, 176; environmental deg-
    radation in, 176, 182–83, 200,
    271n26; food studies in, 182–83;
    and mules, 184, 188–89; and
    pigs, 178–79, 184–86, 187–88,
    195; racism in, 176, 180, 183–84,
    191, 193–94, 197–200, 270n14,
    271n29; and "Uncle Remus"
    stories, 174, 181, 184, 272–73n45;
    and U. S. regionalism, 173, 181–
    83; Washington and Du Bois
    compared to, 180–81, 193
    —Works: "The Conjurer's
    Revenge," 176, 183, 184–91, 195,
    199; *The Conjure Stories*, 17–18,
    172–77, 180–84, 189–90, 191,
    196, 198, 200, 202, 211, 273n47;
    The Conjure Woman, 17, 173,
    174, 175, 179, 190, 191, 200, 201;
    "Goophered Grapevine," 194,
    273n50; "Mars Jeems's Night-
    mare," 175, 189; "Uncle Welling-
    ton's Wives," 177–78; "A Victim
    of Heredity," 176, 183, 191–98,
    274n66
Cheung, Sidney C. H., 277n41
Chicago: and cattle industry, 19,
    122, 204–7, 205, 214, 220, 233,
    235, 274–75n4, 276n21; and
    immigrant communities, 209–
    11, 223
chickens: in animal husbandry, 3,
    24, 59, 62, 81, 98, 109, 122, 238,
    272n33, 272n35; cattle contrasted
    to, 9, 17–18, 71, 120, 172–73,

175, 273n53; in *Conjure Stories*,
176–81, 184, 191–93, 195–200,
273n49; USDA reports on, 179,
270n10

chicken thief, 191–92, 197; racism of perception of, 270n12,
274n66

Chinese Exclusion Act of 1882, 19,
277–78n42

Chinese-Japanese Cook Book (Bosse
and Watanna), 225–26

Chingachgook (Mohican), 56–59;
and relationship to Duncan
Uncas Middleton, 61

Churchill, Winston: review of *The
Jungle*, 209

Chvany, Peter, 166, 268n65

Civil War (U. S.), 9; influence on
U. S. literature, 77, 134, 175–77,
190–91, 199, 202

clams: in *Cape Cod*, 15, 78, 83,
100–105, 107; consumption of,
80, 103, 252n3, 256n64; humans
compared to, 101–2, 109

Clark, William, 1, 4–5, 117, 239n7,
240n8. *See also* Lewis, Meriwether

Clasen, Kelly, 251n49, 251n52, 251–52n59

Cole, Jean Lee, 275n8

colonialism: European, 3, 115,
262n39; settler (U. S.), 6, 10, 24,
128, 135, 138, 158, 235, 262n37.
See also settler colonialism

Colorado: feedlot in Ozeki, 230,
234–35; meatpacking plants in,
236–37; River, 161

Comanche nation, 32

Concord, 79, 82–85, 88, 90, 102,
108, 254n20; farmer, 83; rivers
of, 79, 92

Congress, 47–48; passing Chinese
Exclusion Act, 277–78n42; and
Winnemucca testimony of 1884,
16, 112, 119

conjure, 173, 176, 181–83, 190, 191,
271n23; man, 184–89, 272n35;
woman (Peggy), 191–98

"The Conjurer's Revenge"
(Chesnutt), 176, 183, 184–91,
195, 199

The Conjure Stories (Chesnutt),
17–18, 172–77, 180–84, 191, 196,
198, 200, 202, 273n47; and
Great Migration, 211

The Conjure Woman (Chesnutt), 17,
173, 174, 175, 179, 190, 191, 200,
201

conservation, 14, 52, 56, 58,
68, 86, 251n52; and agriexpansion, 60, 73, 233; and
consumption, 59; and Indian
removal, 116

consumption, 64, 79, 86, 90, 98,
110, 199, 254n17; of beef, 187,
272n38; of Chinese cuisine and
culture, 225–26; of land and
resources, 40, 59, 84, 274–75n4;
of meat, 187, 203, 208, 216; of
moose, 90; and morals, 86;
and production, 15, 82; of rice,
226–27

Contreras, Alicia, 169

Coodley, Lauren, 275n5

Cooper, James Fenimore, 3, 14–15,
53, 75, 77, 86, 215, 233, 250n45;
agri-expansion in, 52, 55, 58, 62,
66, 68, 249n25; and bison, 52,
64–66, 249n35; and cattle, 51–
52, 54–56, 59, 61–63; as father
of Susan, 14, 50, 52, 56, 73; and
Fenimore Farm, 55–56; Irving
compared to, 4, 27, 31, 54, 140,
218, 241n28, 246n73; Native
characters in, 11, 14, 56–58, 60–
61, 63–64, 67–68, 249–50n41;
and race and environment, 53,
56–58, 247n14
—Works: *Last of the Mohicans*, 11,
56, 61, 67; *The Pioneers*, 11, 14,
51–52, 54, 56, 59–60, 63, 68, 76,
247n14, 248n19; *The Prairie*,
14, 31, 52, 54, 59–69, 72, 75–76,
248n21, 249n35, 249–50n41,
250n42
Cooper, Susan Fenimore, 3, 51, 70,
199, 215, 233, 248n19; agri-
expansion in, 55, 60, 69, 71–73,
75–76; and cattle, 54, 59, 71–72,
251n47; and Christian ethics, 76,
252n60; as daughter and amanuensis of James, 14, 50, 52, 56, 73,
76, 246n2; and Fenimore Farm,
55–56; and Iroquois, 69, 74–76,
251n56; and local agriculture,
14–15, 77–78, 250n45, 252n66;
nature writing of, 59–60; and
race and environment, 53–54, 57–
58, 247n14, 251–52n59, 252n66;
Thoreau compared to, 8, 15, 54,

69, 78; visit to local farm, 14–15,
69–72, 251n49, 278n43
—Works: *Rural Hours*, 14, 52, 54,
56, 59–60, 63, 69–77, 70, 78;
"Village Improvement Society,"
77–78
Cooper, William, 55, 71, 247n6
Cooperstown, 15, 54–56, 77; in *The
Pioneers*, 51, 247n14; in *Rural
Hours*, 14, 59–60, 69, 71, 74–76;
significance to U.S. culture, 56;
as Templeton, 51
corn, 11, 24, 50, 63, 218, 228,
272n35; meal, 188, 195
Corps of Discovery Expedition. *See*
Lewis and Clark Expedition
cosmopolitanism, 206–7, 209,
229, 274–75n4
COVID-19, 236–37
cowboy, 17, 25, 47, 231, 238; in
Eaton, 223; in Ruiz de Burton,
139–40, 143, 145, 147, 152–53, 155,
164, 172; in Winnemucca, 116
Crane, Ichabod, 21–23, 49
Crayon Miscellany (Irving), 23
Creek nation, 32–33, 36, 39
critical race studies, 6, 56–57, 182–
83, 236
Cronon, William, 239n2, 274–
75n4, 276n17, 276n21
Cushing, Caleb, 47

Dawes Allotment Act of 1887, 12,
137, 261n24, 263–64n58
Dawson, Melanie, 266n20,
267n47

environment, 2, 6–7, 11–12, 16, 50, 65, 144–45, 206, 231, 256n65, 262n36; agriculture compared to, 19, 54, 56–57, 73, 152, 159, 182–83, 220, 232, 235, 270n8; commodification of, 120, 152, 276n17; degradation of, 14, 19, 52, 57–58, 75, 176, 263–64n58, 271n26; and environmental ethic, 99, 249n25, 251n52, 252n66; heredity contrasted to, 193, 197; and justice, 129, 259n10, 261n31; local, 15, 90, 106, 130, 153; as nature, 6, 35, 64, 105, 116, 164, 199, 216, 218, 236; and race, 6, 10, 53, 57–58, 116, 172, 202; in writing, 144

environmentalists, 7, 86, 116

environmental studies, 6, 115. *See also* ecocriticism

essentializing: through food, 19, 228–29, 233; in race and culture, 198, 204, 206, 227

Eves, Rosalyn Collings, 262n36

expansion, *xvi*, 15–16, 32, 42, 89, 178, 230, 237, 277n41; and agriculture, 1–3, 11–12, 19–20, 29–30, 33–41, 44, 52, 58, 175–78, 232–33; beginnings of, 4–5, 23; cattle and, 8, 14, 19, 45, 112, 115, 135, 137, 177, 228; in Chesnutt, 172, 175–78, 183–85, 199–200, 202, 270n11, 271n26; and colonization, 113–14, 178; Coopers on, 50, 52, 54–55, 57–58, 60, 62–63, 68–69, 71–73, 75–78, 249–50n41; domestic livestock, 9, 22, 40,

50, 83, 100, 105; Eaton on, 206, 220, 224; Irving on, 21–25, 29–37, 39–42, 44–45, 48, 242n10, 245n64; and Jackson, 14, 25, 31, 44, 55; and race and environment, 6, 11, 57, 75, 120, 132, 157, 204, 251n58; in Ruiz de Burton, 157; Sinclair on, 203, 206, 210, 213, 220; in Thoreau, 80, 96, 98, 100, 105, 110; western, 21, 29, 60, 244n49; westward, 6, 9–10, 17, 21–24, 31, 42, 48, 63, 66, 76, 110, 113–14, 119, 123, 157, 177, 184, 199–200, 270n11; Winnemucca against, 111–15, 119–23, 124, 128–29, 131, 133, 135, 137–38, 258–59n7, 263n54

extinction, 9, 243n16

farmers, 2, 9, 12, 25, 42, 82, 233, 241n29, 248n17; and cattle, 11, 14, 20, 141, 152; in Chesnutt, 177–78, 183, 187, 192; Coopers' history as, 55–56, 246n2; Coopers on, 54–55, 58, 64, 71, 74, 76, 77, 249n25, 250n45; Eaton on, 218; Irving on, 21–22, 33, 36–39; in Ruiz de Burton, 141, 152, 172, 265n9, 271n25; Sinclair on, 215–16; in Thoreau, 82–83; in Winnemucca, 125, 258n6

farms, 1, 4, 41, 55–56, 231, 240–41n23; and Cahokia, 11; and cattle, 11, 13, 31, 35, 109, 172, 247n6; in Chesnutt, 177–78, 179, 184, 199, 271n25, 272n35; Eaton on, 218, 223–24, 224; Irving on, 30–

33, 35–37, 39, 49; in Ruiz de Burton, 158, 169–70, 172; S. Cooper on, 14, 56, 60, 69, 71–77, 246n2, 251–52n59; Sinclair on, 215–16; in Thoreau, 82, 109, 253n12; in Winnemucca, 125–26, 137

Federal Meat Inspection Act, 210

Ferens, Dominika, 208, 275n8, 276n16

Ferguson, SallyAnn H., 272–73n45

Fernández-Armesto, Felipe, 7

Finley, James, 256n59

First Indigenous Peoples' Global Consultation on the Right to Food and Food Sovereignty, 135, 263n54

Fiskio, Janet, 12, 112, 121, 262n37

Fitzgerald, Amy J., 240–41n23

food, 36, 49–50, 83–84, 103, 104, 185, 229–30, 232, 235–36, 254n26, 261n24: animals, 3, 18, 23–24, 86, 180, 183, 193, 196, 198–99, 202; cattle or bison, 9, 63–64, 66, 216, 240–41n23; culture and habits, 2, 8, 11, 17, 21, 79–80, 97–98, 192, 200, 225, 227–28, 246n76, 274–75n4; essentializing, 19, 203–5, 208; industrial, 114, 253–54n13; items, 50, 176, 188, 198; justice, 12, 111–12, 130, 135, 172; local or regional, 92–94, 99–101, 104, 108–10, 176, 181, 187, 200, 221, 253–54n3; native or indigenous, 15, 24, 50, 80, 90, 129, 253n10; production, 82, 114, 206, 210, 215, 260n16; reform and safety, 19, 209; security, 236–

37; source, 32, 64, 79, 120–22; sovereignty, 16–17, 110, 112–13, 117–19, 120–21, 123, 125–26, 133, 135–36, 237, 262n37, 263n54; and starvation, 112, 132, 195–96; systems, 110, 112, 237; wild, 23, 98, 103, 105, 253n10

Food and Drug Administration. *See* Pure Food and Drug Act

food studies, 6–8, 112, 115, 182–83, 205, 235–36, 254n17

foodstuff, 2, 3–4, 36, 49, 117, 187, 218, 228

foodways, 123, 129, 262n37; Creek, 33; Paiute, 16, 113–14, 119, 129, 135

Fort Astoria, 40, 42

Fort Gibson, 31

Fort Simcoe, 130

Franklin, Benjamin, 49

Franklin, Wayne, 247n8

Fraser, Evan D. G., 240n18

frontier, 13, 22–23, 50, 221, 230, 248n20, 277n41; and cattle, 7, 9, 59, 80, 232; in Chesnutt, 175, 199; Coopers on, 51, 53–55, 59–61, 63, 66, 68–69, 72, 76–77, 250n42, 250n45; Irving on, 29, 30, 32–33, 36–37, 39, 41–42, 44, 50, 241n2; and Native Americans, 29, 36, 251n58, 262n36; in Ruiz de Burton, 142; in Thoreau, 80–81, 97; Turner's thesis on, 25, 242n10

Gabaccia, Donna, 235–36, 274–75n4

game, 116–17, 126, 203–4; absence of, 132; U. S. agriculture replacing, 24, 58–59, 75, 111–12, 125

129–30, 135; horses in, 119–20, 122, 124–25, 130, 133, 260n16; Natchez, brother of, 126, 127, 133–34, 262n44; Paiute food sovereignty in, 110, 112–16, 118–23, 125–26, 130–31, 133–35, 136, 259n12; Paiutes as cattle in, 130–32; Peabody sisters connected to, 119, 134, 260n18, 263n50; related to Old Winnemucca and Truckee, 118–19, 135–36, 137, 261–62n35; and removal, 112–13, 116, 128, 132, 260n18; and reservation system, 123–26, 127–28, 131–33, 137, 262n36; Ruiz de Burton compared to, 18, 143, 164, 172; settler colonialism in, 128, 135, 136; traditional foods and foodways in, 113–14, 116–20, 122, 125–26, 129, 135; and travel to Washington DC, 119, 126–27, 127, 131–32; and westward expansion, 113–14, 119, 123, 137; Wovoka contrasted to, 136, 263–64n58
—Works: *Baltimore American* article, 132; House Subcommittee testimony, 16, 112, 127–28, 131; letter to Henry Douglas, 121; *Life among the Piutes*, 16, 112, 119, 123–25, 126, 130–31, 133–34, 260n18, 262n44; "The Pah-Utes," 111–12, 116–17, 119–20; *San Francisco Morning Call* article, 125–26, 132
hormones in cattle, 230, 234

horses, 99–100, 124–25, 130, 199, 233, 242n8, 274n70, 276n21; in animal husbandry, 47, 59, 106, 122, 154, 181, 189–90; and cattle, 21, 24, 47, 88, 130, 145; and colonialism, 3, 54; as domestic animals, 3–4, 24, 62, 81, 89, 100, 175, 184; as representative of the West, 21, 47, 63; theft of, 33, 61, 64, 143; value of, in Native communities, 32, 46, 119–20, 133, 260n16
Horsman, Reginald, 6
Howe, LeAnn, 7, 134
Howells, William Dean, 208
Humboldt River, 116, 119
Hurston, Zora Neale, 189

Igler, David, 261n25
immigration, 42, 206, 228; policies against, 19, 225, 275n8, 277–78n42, 278n47
Indenture Act, 146
Indian Princess, 11, 120, 126, 129, 260n20, 261n32
Indian removal, 26–27, 29, 41, 75, 116, 119, 246n74, 262n39; and Indian Removal Act, 3, 31, 80
Indian Territory, 13, 30–32, 36, 39; "Washington Irving, Cattle, and Indian Territory," 21–50
Indian Wars, 4, 75, 112, 125
indigenous food, 15, 80, 90, 129, 253n10, 262n37
industrial agriculture, 120, 129, 136, 205, 209, 215, 240–41n23

industrialization, 9, 77–78, 82–83, 115, 232, 277n41; in agriculture, 11, 19, 83, 86, 129, 178, 205, 220; and agri-expansion, 72, 204; in cattle and livestock production, 23, 163, 209, 248n17; in food systems, 79, 98, 110, 114, 179, 185, 253n13; Sinclair on, 209, 215, 224; in Thoreau, 81–83, 85–86, 253–54n13; in Winnemucca, 120, 129, 136

industrial meat production, 206–7, 230

Inness, Sherry, 277–78n42

Irmschler, Christoph, 250n45

Iroquois nation, 11, 69, 74–76

Irving, Washington, 3, 13–14, 28, 233; absence of cattle in the Northwest in, 40–42; agri-expansion in, 21, 23, 25, 29–30, 33–37, 39–41, 48; assimilation of Creeks in, 32–33, 36, 39; and Astoria expedition, 40, 42–43; biracial farmers in, 36–37, 39; bison in, 25, 31, 32, 33–35, 38, 38, 43, 49, 243n24; cattle in, 21–23, 26–27, 29, 44, 45–50; cattle replacing bison in, 24–25, 30, 32, 36–39; Crows in, 45–46; deer in, 22, 24, 34; frontier in, 22–23, 25, 29–30, 32–33, 36–37, 39, 41–42, 44, 50–51, 242n10; globalization of Pacific Northwest in, 41–42, 244–45n52; horses in, 21, 24, 32–33, 46, 47; and Indian removal, 26–27, 29,

31, 41, 243n16, 246n74; Indian Territory in, 21, 30–32, 36, 39, 243n18; J. Cooper compared to, 4, 27, 50, 54, 66, 140, 218, 241n28, 243n22; King Philip's War in, 22, 24, 26; and Manifest Destiny, 33, 41; on Metacomet (King Philip), 22–24; roman-ticization of Native Americans in, 23, 25–27, 29, 246n73; and western or westward expansion, 21–24, 29, 31–32, 35, 39, 41–42, 44, 45, 48–49, 244n49

—Works: *The Adventures of Captain Bonneville*, 13–14, 23, 29, 40, 44, 45–48, 245n60, 245n64; *Astoria*, 13–14, 23, 29, 40–44, 45, 48, 245n53; *Crayon Miscellany*, 23; "The Legend of Sleepy Hollow," 21–22, 31, 49, 246n76; *The Life of George Washington*, 50; "Philip of Pokanoket," 22–23, 26; *The Sketch Book of Geoffrey Crayon, Gent.*, 21–23, 29, 38, 48–49, 246n76; *A Tour on the Prairies*, 13–14, 23, 29, 30–41, 66; "Traits of Indian Character," 23, 25–26

Iverson, Peter, 258n6, 258–59n7, 263–64n58

Jackson, Andrew, 10, 44; and expansion, 14, 25, 31, 55; and Indian removal, 26–27, 29, 31

Jackson, Helen Hunt, 121, 157

Jacobs, Margaret, 269n72

Mississippian people, 11, 31
Mohegan, John. *See* Chingachgook
monopolization: of agriculture,
123, 163, 268n63; railroad, 144–
45, 148, 152, 163, 168, 268n58,
271n25
Moore, David L., 263n54
moose, 78, 83, 91, 99–100,
254n26, 255n36, 255n42; cattle
contrasted to, 88–89, 95–96,
98; names of natural features,
94–95; as regional cuisine, 15,
80, 84, 87–94, 96–99, 103; story
about whales and, 95–96
Morrison, Ronald, 257n76
mules, 54, 59, 181; in Chesnutt,
175, 184–90, 199, 273n48; in
Hurston, 189
Murphy, Gretchen, 276n14
Murray, Laura, 242–43n15
"My Life in Diet" (Sinclair), 226–27
My Year of Meats (Ozeki), 229, 233,
236

Nabhan, Gary Paul, 114, 258n5
National Association for the
Advancement of Colored People,
115, 200
National Park Service, 7
Native Americans, 45–47, 54–
55, 108, 121, 181, 242–43n15,
243n16; bison compared to, 35,
38–39, 44; characters, 6, 14, 17,
23, 26–27, 29, 57, 68, 75, 172; in
expansion and U.S. agriculture,
11, 27, 48, 53, 116, 129, 250n42;
on Great Plains, 25, 30–32, 60,

63, 67–68, 243n24; and indigenous agriculture, 10, 24, 114,
241n27; removals of, 7, 10, 12,
37, 80; in Ruiz de Burton, 139–
40, 146, 148, 150, 152–53, 155,
157, 164–67, 265n11, 268n58;
wildness or wilderness compared to, 87, 116, 249–50n41; in
Winnemucca, 133–35, 259n14,
259–60n15, 260n18. *See also*
Indian removal; *and individual
nations*
nature, 27, 54, 70, 85, 87, 100, 109;
civilization contrasted to, 26, 62,
77, 93, 223, 265n11; conservation, 56, 86; food and, 7, 84, 108,
110, 235–36, 253–54n13; human
intervention in, 74, 76, 107, 133,
182–83, 248n19, 252n66; Native
Americans compared to, 10, 17;
and race, 57–58, 115–16, 197, 199,
202, 242n10; writing, 12, 14–15,
54–56, 59–60, 68–69, 71, 78
Nauset nation, 101, 105, 108
Negri, Antonio, 40
Nelson, Barbara, 8
Neptune, Governor, 95
Neptune, Louis, 255n48
Nestle, Marion, 237–38
Nevada, 42, 164, 265n13; Paiutes
in, 118, 121, 123, 124, 128, 259n14
Niemeyer, Mark, 242–43n15
noble savage myth, 11, 25, 27, 29,
247n14
North Carolina: in *The Conjure
Stories*, 18, 172–73, 177, 179, 179,
183, 188, 194

post-Reconstruction, 17–18, 172,
183, 190, 202
The Prairie (J. Cooper), 14, 31, 52,
54, 59–69, 72, 75–76, 248n21,
249n35, 249–50n41, 250n42
"The Prairies" (Bryant), 31
Progressive Era, 19, 115
Pure Food and Drug Act, 210

quahog, 84, 102–4, 106
quest: in *Cattle*, 223; for grazing
lands, 115; in *The Maine Woods*,
96; for Paiute sovereignty, 115; in
The Prairie, 61, 63

race studies. *See* critical race studies
racism, 57, 276n14: in Chesnutt,
175–76, 180, 184, 189, 197, 200;
in Ruiz de Burton, 140, 162,
166, 172, 268n65
railroad, 128, 223, 277–78n42;
monopoly, 144–45, 148, 152, 163,
168, 268n58, 268n65; southern,
163, 271n25; wars, 169
Ramirez, Pablo, 264–65n5
Rancho Jamul, *149, 150, 171*
realism, U.S. literary, 160. *See also*
regionalism
Reconstruction: failure of, 12, 172,
177, 180, 183, 190, 202; in literature, 163, 184
Redling, Erik, 272n43
Reed, T. V., 259n14
reform, 120, 209; agricultural, 57
regionalism, U. S. literary, 157, 170,
173, 181, 182–83, 206. *See also*
realism, U.S. literary

reservations, 16, 35, 113, 121,
123–25, 128–29, 132, 258–59n7,
262n36; agents of, 112, 120, 124–
26, 128; Malheur, 128, 261n25;
McDermitt, 127–28; Pyramid
Lake, 128; system of, 125–26, 137,
259n10; Yakama (Yakima), 127,
128, 131, 133
resources, 18, 25, 54, 77, 113, 121,
130, 204, 260n18; allotment of,
116, 122; cattle as competition
for, 16, 22, 119, 123; depletion of,
7, 14, 40, 70; management of,
22, 52, 58, 68, 70, 71, 153, 199
rhetoric, 42, 44, 260n18, 263n50;
of cattle, 111, 114, 120–21, 130,
135; of indigenous knowl-
edge, 118, 263n53; in Ruiz de
Burton, 157–58, 166–67, 170,
269n76; sentimental, 160; in
Winnemucca, 16, 112–13, 117,
120–21, 128, 134–37, 259n12,
260n16
Richardson, Robert, 92–93, 253n12
Rimas, Andrew, 240n18
Rivera, John-Michael, 264n2
Robinson, Kenneth Allen, 253n10
Rodier, Katharine, 263n50
Rogin, Michael, 242n8
Roosevelt, Theodore, 86, 209–10
Root, Waverly, 31, 36, 244n34,
272n33, 273–74n60
Ruiz de Burton, María Amparo, 3,
16, *149*, 202, 233; and Califor-
nios/as, 140–46, 148, 151–54,
156–69, 265n9, 265n13; cattle in,
17, 139–46, 148, 152–54, 156–61,

Ruiz de Burton (*cont.*)
163–66, 169–72, *171*; Chapo in, 165–67, 222, 268n65; cowboys in, 143, 145, 147, 152–53, 155, 164, 172, 265n13; Kumeyaay characters in, 140–41, 143–46, 148–50, 152–53, 155–70, 172, 266n26; and Mexican-American War, 142, 145, 170; race and ethnicity in, 144, 155, 157, 165, 169; railroad monopolists in, 163–64, 271n25; settler colonialism in, 151, 158; squatters in, 139–46, 148, *149*, 152–53, 155, 157–59, 161–64, 166–69, 266n20; Treaty of Guadalupe Hidalgo in, 16–17, 138, 140, 142–43, 151, 158, 164; vaqueros in, 17, 139–40, 143–48, 151–52, 154–57, 159–62, 165–67, 169, 172, 265n13; Winnemucca compared to, 18, 137–38, 143, 164, 172
—Works: *Letters*, 169–70; *The Squatter and the Don*, 16–17, 138, 139–72, *171*, 264n2, 268n53
Rural Hours (S. Cooper), 14, 52, 54, 56, 59–60, 63, 69–78, *70*, 250n45, 251n52

Sacajawea, 134
Saldivar, José, 143, 152
Sánchez, Rosaura, 169–70, 269n75
San Diego, 17, 140, 148, 162, 163, 168, 264n1, 266n21
San Francisco Morning Call article (Winnemucca), 125–26, 132
Schachterle, Lance, 248n22

Scherer, Joanna Cohen, 261n32
Schlosser, Eric, 229, 276n17
Schlueter, John, 38
Schnell, Michael, 246n76
Serra, Father Junipero, 47; mission system, 145
settler colonialism, 6, 10, 24, 57, 128, 135, 151, 158, 228, 235, 250n42, 262n37. *See also* colonialism
Shapiro, Laura, 225, 278n50
sheep, 3, 59, 82, 106, 124, 238; Coopers on, 54–56, 62–64, 73, 251n47; Sinclair on, 212, 276n21; in Thoreau, 89, 96, 98. *See also* goats
Sheridan, Philip, 10
shipwreck of *St. John*, 100, 104–5, 256nn63–64
Shour, Nancy, 247n9, 248n19
Sierra Club, 115
Sierra Nevada, 159–60
Simon, Katie, 256n65
Sinclair, Upton, 3, 12, 18–19, 163, 203–7, 220, 223–24, 229, 233–34, 275n5; and immigration, 209–11, 214–15, 278n47; and meat production, 9, 202, 211–16; and rice, 225–28
—Works: *The Jungle*, 9, 13, 18–19, 203–7, 209–16, 229, 234, 241n31; "My Life in Diet," 226–27
Sioux nation, 5, 29, 214; in *The Prairie*, 60–64, 67–68; as Sieux, 5
Sivils, Matthew, 249n25
Sizemore, Michelle, 243n16

Treaty of Wanghia of 1844, 48
tribalography, 16, 134
Truckee, 116, 118–19, 137, 260n18,
261–62n35
tuberculosis, 109, 253n10
Turner, Frederick Jackson, 13, 25,
54, 242n10, 247n6
Tuskegee Institute, 181; and beef,
18, 180, 193
Tuttle, Jennifer, 268n53
Twain, Mark, 49, 258n89
Tyler, Alice Felt, 277–78n42

"Uncle Wellington's Wives"
(Chesnutt), 177–78
Up from Slavery (Washington), 193
urbanization, 9, 77, 206
USDA, 179, 270n10

Vallejo, Mariano Guadalupe, 150
Van Ballenberghe, Victor, 90,
255n36
Vance, William, 64, 249n31,
249n35
vaqueros: Irving on, 45, 47; in Ruiz
de Burton, 17, 139–40, 143–46,
148, 151–57, 159–62, 165–69, 172,
265n13
"A Victim of Heredity" (Chesnutt),
176, 183, 191–98, 274n66
"Village Improvement Society" (S.
Cooper), 77–78
violence: in Chesnutt, 200; Coopers on, 251n58; Eaton on, 218;
Irving on, 45; in Ruiz de Burton,
153–54
Vizenor, Gerald, 263n52

Wabanaki hunters, 94
Wagner, Frederick, 81
Walden (Thoreau), 69, 78, 86, 94,
109; cattle in, 82; woodchuck in,
15, 79, 83
Walker, Cheryl, 263n50
Wallach, Jennifer Jensen, 181, 193,
272n38, 274n71
Walls, Laura Dassow, 83, 109–10,
253n12, 258n90
Wampanoag nation, 10, 22, 26–27,
101, 242n8
Warford, Elisa, 267n41
War of 1812, 27, 40
Warren, Louis S., 263–64n58
Washington, Booker T., 17–18, 180
Washington, George, 29, 50, 240n23
Watson, David, 44
A Week on the Concord and Merrimack Rivers (Thoreau), 15, 81–82,
84–87, 92, 100, 105, 110
Weinstein, Josh, 252n60
Welfleet Oysterman, 108, 252–
53n3, 256n64
the West, 12–16, 172, 177, 219, 235,
241n29, 261n23, 272n35; and
animal husbandry, 23, 176; in
Chesnutt, 179, 189; Coopers
on, 54, 62, 69; cowboys in, 147,
157; domestic animals of, 5, 36,
63; Eaton on, 209, 219; and
expansion, 1, 18, 33, 81, 110, 172;
extension of slavery in, 18; George
Gatlin paintings of, 38; Irving on,
23, 33, 40–42, 46–48, 245n64;
in Ruiz de Burton, 142, 157; in
Winnemucca, 121, 123, 138, 259n10

western expansion, 21, 29, 73, 244n49. *See also* westward expansion
westward expansion, 6, 9, 270n11; in Chesnutt, 177, 184, 199; Coopers on, 62–63, 66, 76; Irving on, 21–24, 30–31, 42, 48; monoculture in, 17; in Ruiz de Burton, 157; and slavery, 10; in Winnemucca, 110, 113–14, 119, 123. *See also* western expansion
wheat, 63; in Chesnutt, 197; Eaton on, 218, 228; in Ruiz de Burton, 142–43, 148, 158, 167, 169
whiteness, 58, 274n71
Whitman, Walt, 235
Wilbur, James H., 130–31
wilderness, 11, 57, 80, 88, 98, 164, 242n10, 257n76; conservation of, 86, 129, 250n45, 259n10; domestic plants and animals in, 96, 105, 222, 244n49; as doomed or conquered, 35, 176, 248n19; Native characters in, 11, 27, 90, 93, 116
wildness, 32, 96, 222; and bison, 24, 33–34, 66, 243n24; civilization contrasted to, 31, 99, 105; and moose or clam, 83, 86, 90; Native Americans in, 35, 87

Williams-Forson, Psyche, 180, 192, 270n12
Williamson, William, 91
Wilson, Jack. *See* Wovoka
Winant, Howard, 7, 57, 136, 183
Winnemucca, Chief, 127, 135, 137
Winnemucca, Natchez, 126, 127, 133–34, 262n44, 263n50
Winnemucca, Sarah. *See* Hopkins, Sarah Winnemucca
Winthrop, John, 115, 228
Wolfe, Cary, 236, 262n39
World Food Summit, 135
Wovoka (Jack Wilson), 136, 263–64n58
Wu, David Y. H., 277n41

Yakama reservation, 127–28, 131–33, 137; as Yakima, 127, 131, 133
Ybarra, Priscilla, 144, 172, 265n11, 267n44
Yellowstone National Park: and bison territory, 66, 233; and indigenous communities, 6, 9, 129

Zanjani, Sally, 118, 134, 135–36, 261n24

To order or obtain more information on these or other University
of Nebraska Press titles, visit nebraskapress.unl.edu.

Lightning Source UK Ltd.
Milton Keynes UK
UKHW010252290421
382821UK00001B/25

9 781496 218643